Turing Tales

Edgar G. Daylight
Contributions by Arthur C. Fleck and Raymond T. Boute

Edited by Kurt De Grave.

LONELY SCHOLAR™
SCIENTIFIC BOOKS

First edition
Version 1.0

© 2016 Edgar G. Daylight
Daylight can be contacted at egdaylight@dijkstrascry.com.

Published by Lonely Scholar bvba
Dr. Van De Perrestraat 127
2440 Geel
Belgium
http://www.lonelyscholar.com

Typeset in LaTeX

D/2016/12.695/1
ISBN 978-94-91386-06-0
NUR 980, 686

For Alida, Helena & Ben

Contents

Preface

Yesterday I received the following absurdity about full formal verification in my mailbox. The italicized words are particularly frustrating.

> Message: 3 Date: Tue, 27 Dec 2016 10:09:17 -0500
> From: ...
> To: ...
> Subject: ... DeepSpec Summer School, July 13-28, 2017
> Message-ID: <...@cis.upenn.edu>
> Content-Type: ...
>
> The first DeepSpec Summer School on Verified Systems will be held in Philadelphia from July 17 to 28, 2017, preceded by an introductory Coq Intensive from July 13 to 15.
>
> Overview
> *Can critical systems be built with no bugs whatsoever in hardware, operating systems, compilers, crypto, or other key components? It may seem a pipe dream, but in fact the past decade has seen explosive advances in the technology required to realize it.*
>
> This summer school aims to give participants a wide-ranging overview of several ambitious projects currently underway in this space.
>
> Participants will complete the summer school with a thorough understanding of the conceptual underpinnings of these projects plus considerable hands-on experience with state-of-the-art tools for building verified systems. ...

The present book, by contrast, will provide you with some understanding of why full formal verification is actually *not* possible: not in principle and thus not in practice either.

That being said, I am an advocate of formal methods and today I came across the following informative text from the Trustworthy Systems Research Group @Data61:

> With all the purity and strength of mathematical proof it is easy to get carried away. There is a fundamental theoretical limit of formal verification: there will always be some bottom level of assumptions about the physical world left and these assumptions have to be validated by other means. Mathematical proof is proof as long as it talks about formal concepts. It is when assumptions connect to reality where things may still fail. Albert Einstein is quoted as saying "As far as the laws of mathematics refer to reality, they are not certain; and as far as they are certain, they do not refer to reality." For instance, if the hardware is faulty, or if a cosmic ray randomly changes memory, the correctness predictions of our proof do not apply. Neither do any traditional tests or verification methods help against cosmic rays, but it is important to keep in mind that *even with mathematical proof, there are no absolute guarantees about the physical world.*
> [Source: *http://ts.data61.csiro.au/projects/TS/l4.verified/proof.pml*]

This statement is illuminating but only lifts the tip of the veil. Guaranteeing the absolute correctness of an industrial digital system is intrinsically hard (to say the least) because of multiple reasons, two of which are:

1. Many, if not most, errors in software engineering occur when bridging the gap between the informal, real world and the formal world of mathematical specifications. This is what Edsger Dijkstra misleadingly called the "Pleasantness Problem" — a topic that I have touched on in almost all my oral histories and especially with Peter Naur and Michael Jackson (see: *www.lonelyscholar.com*).

2. Even if software engineers have a clear-cut specification of how they intend their software to behave, they will at best be able to prove that a mathematical model of their software satisfies this specification, not that the software will have the desired effects in the real world.

It is the second argument that the Research Group @Data61 alludes to, but not the first. Excellent books that cover the limitations of formal verification are: Timothy Colburn's *Philosophy and Computer Science* (2000), Donald MacKenzie's *Mechanizing Proof: Computing, Risk, and Trust* (2004), and Nancy Leveson's *Engineering a Safer World: Systems Thinking Applied to Safety* (2012).

Furthermore, I will stress in the present volume that not one but several mathematical models of the physical world can — and should — be used to advance engineering. For example, I can model my laptop with both a finite state machine and a linear bounded automaton without contradicting myself. And a similar remark can be made about the computer programs residing in my laptop. My point is that the strength of mathematical proof is *not only* limited from below because of "some bottom level of assumptions," as mentioned in the previous excerpt. Restrictions also follow from the limitations that each mathematical model has individually. The best one can do is prove properties of *a* mathematical model on day one, and prove properties of *another* mathematical model on day two, and so on. By the end of the week, the software engineer will have obtained more confidence (but definitely no guarantee) about the correctness of his or her digital system. There is no such thing as the most adequate mathematical model of the engineering artefact at hand — at least not for the many industrial case studies that are discussed and mentioned in this book.

From a historical perspective, I shall argue that some prominent researchers in computer science have not consistently distinguished, and still do not consistently distinguish, between some of their favorite abstract objects (such as a universal Turing machine) and the representations thereof in the real world. Believe it or not, a universal Turing machine is often used as a synonym for a computer; that is, researchers often forget that the former is merely a mathematical model of the latter. Not seeing this triviality and others, in turn, leads to flawed reasoning and category mistakes. One such category mistake is claiming that full formal verification is possible.

My historiographical findings will also reveal that the connection between universal Turing machines and computers was only made gradually in the post-war years and especially in the 1950s. Moreover, the people who were making these connections were very few in number. The crux, then, is that, contrary to public opinion today, most computer pioneers were *not* trying to build a practical realization of a universal Turing machine. Such historical nuance will hopefully result in more informed generations of computer scientists in the decades to come.

Finally, it is with gratitude that I refer to the contributions of Arthur Fleck and Raymond Boute. Fleck's chapter contains an extensive personal recollection on the history of programming languages. Boute's technical scrutiny pertains to unsound treatments regarding the *function* concept in mathematics, with the conclusion that in "most *introduction to proof* texts one has to wait for over 100 pages, only to be disappointed by flawed definitions."

Provocations abound! I look forward to receiving feedback from various scholars, including the attendants of the aforementioned DeepSpec Summer School on Verified Systems.

— Geel, Belgium, December 29, 2016

1. Introduction

Researchers from computer science, linguistics, physics, and other disciplines have a tendency to speak about their mathematical models as if they coincide with reality. The following statement, for example, is not uncommon in physics:

> Kurt Gödel proved that time travel is possible.

Careful researchers, by contrast, will stress at least once in their writings that:

> Kurt Gödel proved that time travel is in principle possible *according to his mathematical model*.

More sentences of the first kind are presented in this book with regard to computer science. They show that researchers frequently fuse their mathematical models with their engineered systems.

Mistaking the category of mathematical objects for the category of concrete physical objects amounts to making a *category mistake*. Some category mistakes are less innocent than others, as Chapter 6 — A Titanic Turing Tale — will reveal. For example, it is easy to find computer science papers in which the authors explicitly and incorrectly state that they have mathematically *proved* that certain *systems* cannot be *engineered*. While many people would like computer science *theory* to take precedence over software *engineering*, nobody wants this ideal if the price to pay is faulty reasoning. Rectifications are required and as a brief indication of where I am heading, Sections 1.2 and 1.3 in the present introduction contain my scrutiny of statements made by Michael Hicks and referees of computer science's flagship journal, the Communications of the ACM.

Category mistakes are discussed at length in the second part of this book, while the first part reveals another tendency in computer science: that of constructing a set of foundations in which ideas that had not been understood as connected when they were put forward are retrospectively integrated. Later generations of computer scientists then assume that these things have always been related. The connection between universal Turing machines and computers is one notable example. Many computer scientists who have given invited talks about the history of their discipline have made statements of the following kind:

> During the 1950s, a universal Turing machine became widely accepted as a conceptual abstraction of a computer.

Other scholars, by contrast, will refrain from making such general and appealing claims. Instead, they will focus on specific actors and write the following, for example:

> By 1955, Saul Gorn viewed a universal Turing machine as a conceptual abstraction of his computer.

The subject of the first sentence is not a historical actor, unlike the subject of the second sentence. If you happen to prefer the first sentence and tend to write sentences of this kind, then it is likely that your exposition will, at best, capture a development of ideas that is detached from the people who shaped the discipline under investigation. As a result, neither you nor your readership will realize that a universal Turing machine had different meanings for different actors, nor will it become apparent that the meaning of a Turing machine changed over time for each individual actor.

We will learn more about Saul Gorn and Turing machines in Chapter 2: A Tatty Turing Tale. The word "tatty" is a synonym for "worn out" and it is indeed a worn-out tale — at least for historians today — that Turing supposedly invented the modern computer. It is likely that you know somebody who proclaims that Turing is the inventor of the computer because his *1936 theory* was a prerequisite for later breakthroughs in *engineering*. Chapter 2 reveals that this romantic view of scientific progress is misleading to say the least. Chapter 3 illustrates that there is a Dutch exception to the rule.[1]

The overarching theme of the present book, then, is the relationship between theory and engineering. Two specific topics are the histo-riographical mistakes and category mistakes made by researchers of standing in their quest for a utopian world, a world in which the role of mathematics in engineering is intended to be that of a queen rather than a humble servant.[2]

My meta-challenge is to convince the reader that thorough historical and philosophical investigations into computer science can advance computer science itself. As I will argue in this chapter, there is a marked difference between the computer scientist of today and the computer scientist of tomorrow.

1.1   John Reynolds and Turing's 1936 Paper

In order to adhere to academic standards, Gerard Alberts, a historian of mathematics and computing, has advised me to avoid using technological and theoretical concepts such as 'program,' 'compiler,' and 'universal Turing machine,' as the subjects of my sentences. Disregarding his advice often amounts to writing expositions in which one line of thought dominates the entire story. For example, if I choose as a subject matter Turing's 1936 paper [320], or more specifically, the connection between Turing's 1936 universal machine and modern computers, then it becomes very tempting to view the history of computing in terms of those few people who grasped that connection early on. Turing, for example, already saw a connection around 1944. This observation alone, however, does not make him an influential actor in the history of science and technology. By following Alberts's advice, the scholar loses the inclination and the temptation to paint the history of computing as a continuous stream from Turing's 1936 paper to the stored-program computer. The scholar will fail in explaining every advancement in terms of Turing machines, and this is to the good.

The late Michael Mahoney warned his fellow historians not to fall into the trap of viewing everything with the same glasses. The computer, Mahoney said specifically, is not one thing but many different things. He continued as follows:

> [T]he same holds true of computing. There is about both terms a descriptive singularity to which we fall victim when, as is now common, we prematurely unite its multiple historical sources into a single stream, treating Charles Babbage's analytical engine and George Boole's algebra of thought as if they were conceptually related by something other than twentieth-century hindsight. [234, p.25-26]

In my own words, then, the multitude of computer-building styles, programming habits, receptions of Turing's work, interpretations of the terms "recursion" and "computer program," and so on should be placed front and center by computer scientists and historians alike. Terms such as "general purpose" and "Turing universal" had different meanings for different actors, and their usage as synonyms is only justified if this conforms with the historical context. I take Mahoney's challenge to mean that we need to handle each and every term with great care.

Unfortunately, Mahoney "never took up his own invitation," says Thomas Haigh in the introduction of Mahoney's collected works [234, p.5,8]. Men like John McCarthy, Dana Scott, Christopher Strachey, Turing, and John von Neumann appear in almost every chapter in Mahoney's collected works, but, as Haigh continued,

> [W]e never really learn who these people are, how their ideas were formed and around what kind of practice, what their broader agendas were, or whether anything they said had a direct influence on the subsequent development of programming work. [234, p.8]

Mahoney did occasionally provide some insights about the aforementioned men. Coincidentally or not, the rare passages in which he did also comply with the writing style advocated by Alberts, as the following excerpt from Mahoney's oeuvre illustrates.

> Christopher Strachey had learned about the [lambda]-calculus from Roger Penrose in 1958 and had engaged Peter J. Landin to look into its application to formal semantics. [234, p.172]

Words such as these inform the reader about several historical actors, their social networks, and their research objectives.

In general, however, Mahoney much preferred technological concepts over historical actors.[3] One instructive example comes from his 2002 work, in which he stated that:

> The idea of *programs that write programs* is inherent in the concept of the universal Turing machine, set forth by Alan M. Turing in 1936. [234, p.78-79, my emphasis]

This statement, which has an idea as subject matter, is anachronistic. Turing's 1936 paper was not at all about programs that write programs. Turing's universal machine did not modify its own instructions, nor did it modify the instructions of the other machines that it simulated.

Instead of alluding to compilers, as Mahoney did, one could perhaps refer to interpreters when attempting to document the importance of

Turing's 1936 paper. After all, interpreters emulate programs — loosely speaking. The following sentence is technically more accurate:

> [Turing's] universal machine in particular is the first example of an interpretative program. [88, p.165]

These words come from the eminent logician Martin Davis who has made great strides in explaining to the layman the importance of logic is in computer science. The subject in Davis's sentence is Turing's universal machine, not a historical actor. From Alberts's perspective, then, it is tempting to rewrite, and in my opinion improve, the wording. My first suggestion is to write:

> During the early 1960s, John McCarthy viewed a universal Turing machine as an interpretative program.

This sentence limits the scope of Turing's influence by focusing on McCarthy and the years in which he thought about interpretative programs (in the modern sense or in *a* more modern sense of the word).

My second, more elaborate, suggestion is to write:

> In the 1950s, when interpreters were built, leading computer programmers like McCarthy did not initially view Turing's universal machine as an interpretative program in a practical sense. Although McCarthy had by 1960 already written a paper [238] in which he had connected the universal Turing machine to his LISP programming system, he did not see the practical implication of LISP's universal function *eval*. It was his student Steve Russell who insisted on implementing it. After having done so, an interpreter for LISP "surprisingly" and "suddenly appeared".[4]

The previous fragment informs the reader about McCarthy and his research team, partially capturing how Turing's 1936 universal machine eventually led to McCarthy's interpreter for LISP. The passage does not disregard the possibility that others had already made some kind of a connection between a universal Turing machine and programming technology in earlier years. However, if one wants to make a general claim about Turing's legacy, then he or she should first do the hard work of finding several specific actors for which the claim holds.

Taken at face value then, Mahoney's oeuvre gives the false impression that several influential historical actors, along with himself, thoroughly understood Turing's 1936 paper and the logic literature in general. "Computer science," Mahoney postulated in 1997, "had come around full circle to the mathematical logic in which it had originated" [234, p.144]. However, this claim does not sit well with his own appeal to avoid falling into the trap of viewing everything with the same — and in this case, logical — glasses time and again. Nor does it sit well with several primary sources and oral histories. For example, according to an expert in Separation Logic, the late John Reynolds, understanding the logical literature is "taxing."

> [Reynolds:] I'd better admit that I haven't read Turing's 1936 paper. I probably avoid old papers less than most computer scientists, but I wasn't trained in logic and thus find the subject taxing to read (indeed, more to read than to write). [288]

Men of the stature of Tony Hoare and Peter Naur had difficulty studying Turing's 1936 paper [98]. In addition, Davis's remark at the end of Christos Papadimitriou's talk at Princeton in 2012 clearly shows that prominent computer scientists, such as Papadimitriou, still do not fully comprehend Turing's 1936 paper either [270].

In his monumental book *A Science of Operations* [280], Mark Priestley documented the interaction between theory and practice in the development of computing, starting from the early work of Charles Babbage and ending with the programming language Smalltalk, thereby providing a coverage that has yet to be matched by fellow historians. Based on his work and on some of my own research, I now list seven influential publications of the 1950s-1960s. Each publication played an important role in transferring ideas from logic to computing.

- 1950: Turing's *Computing machinery and intelligence [322]*

- 1950: Paul Rosenbloom's *Elements of Mathematical Logic [294]*

- 1952: Stephen Kleene's *Introduction to Metamathematics [202]*

- 1954: Andrey Markov Jr.'s *Theory of Algorithms [197]*

- 1958: Davis's *Computability and Unsolvability [86]*

- 1958: Haskell Curry & Robert Feys's *Combinatory Logic, Vol. 1 [78]*

- ...

- 1967: Marvin Minsky's *Computation: Finite and Infinite Machines* [245]

The influence exerted by most of the publications listed above has yet to be thoroughly investigated in future work. Turing's scholarly legacy in computer science, not to mention that of Emil Post, Alonzo Church and others, has yet to be described.

Priestley has discussed the influence of the first publication at length in 'Logic and the Invention of the Computer' [280, Ch.6]. Priestley argued that Turing had little influence in the 1940s (and in computer building in particular) but that his 1950 paper, 'Computing machinery and intelligence,' was "a turning point in the characterization of the computer as a universal machine" [280, p.153]. "After 1950," he wrote, "it became common to describe electronic digital computers as being instantiations of Turing's concept of a universal machine" [280, p.124]. Furthermore, he writes:

> Following 1950, [Turing's 1950] paper was widely cited, and his characterization accepted and put into circulation. [280, p.153]

These words by Priestley give the unfounded impression that most computer practitioners became acquainted with Turing's work during the 1950s. If I now name people who did not grasp, read, or come across Turing's work, Priestley's readership will think that these are exceptions. By the late 1950s, Turing's work had indeed become increasingly popular in some niches of computing (see Chapters 2 and 3). However, if anything at all can be stated about the majority of computer practitioners, then it is that they either did not yet understand the all-purpose nature of the computer, or they did but *without* having resorted to (a re-cast version of) Turing's work [98]. Ideally, and following Alberts, neither Priestley nor I should be making claims about the majority of computer practitioners in the first place. Instead of taking Turing's 1950 paper as a subject, as Priestley has done in the above passage, a historical actor or a well-defined group of historical actors could be chosen instead, thereby limiting the scope of Priestley's claim regarding Turing's influence.[5]

To conclude, then, I urge computer scientists and historians to respect the various receptions of Turing's work. Furthermore, let us view our past not solely as an application of Turing's 1936 paper but also as a history of struggling to understand what Turing's 1936 paper has to offer to some, perhaps many, but definitely not all of us. This brings me to my final introductory remarks about Chapters 2 and 3, chapters which

I wrote with the previously explained methodology and motivation in mind. Both chapters cover the 1950s and the role that Turing's 1936 paper played in the emerging discipline now called computer science. Chapter 2 focuses on Anglo-Saxon developments and Chapter 3 zooms in on the Dutch cities Delft, The Hague, and Amsterdam. It was in Delft and later in The Hague where Turing's 1936 paper *did* influence computer *building* to an exceptional extent, both on an academic level and in an industrial European-wide context due to the extraordinary work of Willem van der Poel.

My main interests in Chapters 2 and 3 are, again, the relationship between modern logic and engineering and the way in which this relationship is erroneously portrayed today, albeit often unintentionally, in order to advance a particular research agenda.

1.2 Michael Hicks and Formal Verification

Viewing a universal Turing machine and a stored-program computer as one and the same amounts to making a trivial, yet common, category mistake. Jack Copeland is perhaps the most renowned scholar who insists that Alan Turing came up with "the stored-program universal computer" as a "single invention" in 1936. Andrew Hodges, Martin Davis, and others have, in turn, complained about Copeland's viewpoint and I suspect that Turing would do the same if he were still alive today.[6]

Why would Turing, of all people, mistakenly view his mathematical *model* of computation, now called a Turing machine, as a concrete physical object such as a stored-program computer? A Turing machine can be practically realized in multiple ways, and a stored-program computer is just one of them. Furthermore, and as I will highlight repeatedly, a Turing machine is only one possible model of a computer.

Category mistakes are not limited to digital systems, however. Engineers in all areas use models and can potentially fall into the trap of mistaking the model for reality. Dave Parnas told me that one of his favorite examples in this regard is Ohm's Law, which states that $V = I \times R$. In reality, the ratio between current (I) and voltage (V) varies, i.e., the relationship is not actually linear. Engineers can use this law to analyze a circuit and verify that it has the desired behavior only to find that the circuit fails in certain conditions. Parnas gives the following example:

> The circuit may overheat and the resistance (R) will change. Ohm's Law can be made an accurate description of the device

by adding conditions, e.g. $V - I \times R < \delta$ if $a < I < b$ etc. The result will be much more complex, but it will give conditions under which the model is accurate. Even this is a lie, as it makes assumptions about ambient temperature and the aging of the components that are not stated.[7]

Parnas's main point is that engineers must be aware of the limitations of models as descriptions. More on Parnas's views can be found in the proceedings of *The Future of Software Engineering*, a workshop held in ETH Zurich in November 2010 [103, 255]. Parnas concludes by noting that this awareness is completely missing from the UML (Unified Modeling Language) and MDE (Model-Driven Engineering) literature. Can the same be said about the programming language literature? The full answer to this question is given in Chapter 6.

Of course, one could argue that researchers frequently choose to craft their sentences with brevity in mind and that they are well aware of the fact that they are, strictly speaking, making small category mistakes; surely, they must know that these are models. This is precisely the kind of response I received from a POPL referee a few months ago with regard to my rejected paper, entitled: 'Category Mistakes in Computer Science at Large.' (POPL is an abbreviation for Principles of Programming Languages and the contents of the aforementioned paper constitute Chapter 6.) When I gave this kind of response to the colleague who brought to my attention the statement about Gödel and time travel, he insisted that some physicists really do believe that Gödel proved that time travel is possible. As this book will reveal to the neutral scholar, the same can be said about a large number of computer scientists.

Similar remarks were made by the mathematician Vladimir I. Arnold in 1997:

> The mathematical technique of modeling consists of [...] speaking about your deductive model in such a way as if it coincided with reality. The fact that this path, which is obviously incorrect from the point of view of natural science, often leads to useful results in physics is called "the inconceivable effectiveness of mathematics in natural sciences" (or "the Wigner principle"). [15, my emphasis]

The mathematical technique of modeling is undoubtedly useful in practice. However, society will fare even better when the vices of modeling are also placed front and center every now and then.

Coming to formal verification, then, several prominent computer scientists — some of whom I have met and talked to many times — actually believe the following statement:

> ... full formal verification, the end result of which is a proof that the code will always behave as it should. Such an approach is being increasingly viewed as viable. [176]

These words come from Michael Hicks, who said in 2014 that we can have a mathematical *proof* of the behavior of an engineered *system*. The best we can actually do — as already pointed out by Brian Cantwell Smith [303], James Fetzer [125], and Timothy Colburn [72] in previous decades and in a technically more refined way in the present book — is to prove certain *mathematical properties* of one or more *mathematical models* of the running *system*. This small nuance will presumably be readily accepted by many readers, but I hope to bring more conceptual clarity to the table in the sequel.[8]

Rice's Theorem Under Attack

Let us consider some of the alleged practical implications of Rice's theorem. Here, my own thoughts become more visible, since the writings of Fetzer and the other two aforementioned philosophers do not cover computability theory.

For the uninitiated, Figure 1.1 presents Wikipedia's description of Rice's theorem. For both the uninitiated and the initiated, Figure 1.1 comes very close to illustrating a category mistake, and therefore exemplifies a lack of conceptual clarity that I hope to remedy in the present book. I will also return to Figure 1.1 in the final chapter.

Stepping away from Wikipedia for now and coming to Hicks's discussion on software bugs and Rice's theorem, this is how Hicks defines a "sound" mathematical analysis:

> A *sound* analysis is one that, if there exists an execution that manifests a bug at run-time, then the analysis will report the bug. [176]

Note that the first occurrence of the word "bug" refers to an event in the real, physical world. The second instance of the same word refers to a mathematical model of a bug (and only indirectly to the manifested bug in the real world). For the sake of clarity — and at the expense of some

> In computability theory, Rice's theorem states that all non-trivial, semantic properties of programs are undecidable. A semantic property is one about the program's behavior (for instance, does the program terminate for all inputs), unlike a syntactic property (for instance, does the program contain an if-then-else statement). A property is non-trivial if it is neither true for every program, nor for no program.
>
> We can also put Rice's theorem in terms of functions: for any non-trivial property of partial functions, no general and effective method can decide whether an algorithm computes a partial function with that property. Here, a property of partial functions is called trivial if it holds for all partial computable functions or for none, and an effective decision method is called general if it decides correctly for every algorithm.

Figure 1.1: Wikipedia's description of Rice's theorem portrays rather accurately how it is taught in academia today in the sense that the first paragraph comes very close to illustrating a category mistake. (How close depends on how careless one answers the question: What, precisely, is a program?) This mistake is discussed at length in the present book. The second paragraph is less problematic but is not very accessible to the average computer scientist for it requires prior knowledge of partially computable functions and computability theory in general. For the record: I accessed the Wikipedia site on October 26, 2016.

brevity — it is worthwhile to make this categorical distinction explicit *at least once*, as I am doing here and will do throughout most of this book.

Another important remark is that if the first occurrence of "bug" indeed refers to an event in the real world, then so does the word "execution." However, the mathematical analysis has to rely on a mathematical notion of execution. A distinction is therefore required between a 'real execution' and a 'mathematical execution.' The former belongs to the category of concrete physical objects, while the latter belongs to the category of abstract objects.

Furthermore, Hicks's 'mathematical execution' denotes a mathematical object in conformance with his chosen model of computation. Somebody else — following, say, Edsger Dijkstra's or Parnas's schools of thought, described below — might choose *another* model of computation. So it would be clearer if Hicks would also make his chosen model of computation explicit in his definition.

In other words, a sound analysis says something about a mathematical model and only *indirectly* something about the computer program that is being modeled. Furthermore, since someone else can choose another model for the same computer program, **it is misleading to suggest that there is a one-to-one interdependence between the computer program and the chosen mathematical model.**

Similar remarks can be made about Hicks's definition of a "complete" analysis, which is as follows:

> On the flip side, a *complete* analysis is one that, if it reports a bug, then that bug will surely manifest at run-time. [176]

Again, a mathematically modeled bug (which the analysis "reports") is categorically distinct from a bug encountered during the "run-time" execution of a computer program. They are not the same thing.

Am I right to conclude that Hicks and other programming language specialists often think they are *directly* referring to *both* their mathematical model *and* the actual computer program? (Chapters 5 and 6 will delve into these matters further.) Later on, Hicks provides the following statement, which I, as a 'POPL outsider,' have genuine difficulty in comprehending:

> Ideally, we would have an analysis that is both sound and complete [Yes], so that it reports all true bugs, and nothing else **[No!]**. [176]

This statement can only be correct if the category of abstract objects coincides with the category of concrete physical objects. However, since a bug encountered during program execution is categorically different from a mathematical bug, what does Hicks's sound and complete analysis *actually* report? Does it report

1. bugs manifested during program execution,

2. mathematical bugs, or

3. both?

Again, I claim that the correct answer is 2. Hicks's analysis reports about mathematical bugs, that is, mathematically modeled bugs. Does Hicks think the answer is 3., in the sense that both categories coincide? Or am I wrong to question whether there is a one-to-one interdependence between the computer program and a well-chosen mathematical model of the computer program?

To recapitulate, at least as far as my understanding goes: a perfect *mathematical* analysis (in the sense that it is both sound and complete) cannot *guarantee* something about the *real* world, including the behavior of the engineered *system* under scrutiny along with the bugs that manifest themselves in the process. This would only be possible if the mathematical object and the engineered system belong to the same category (and, *moreover*, if we can specify absolutely everything about the engineered system in a concise and useful manner). A sound and complete analysis can provide engineers with *extra confidence* that their system will behave appropriately in the real world and *nothing more*.

To continue with Hicks's views on formal verification, it should also be mentioned at this point that, contrary to Hicks, computer scientists following Dijkstra and Adriaan van Wijngaarden would mathematically model a computer program — such as a C computer program — with a Turing-incomplete model of computation and, specifically, with a finite state machine.[9] I therefore *also* struggle with what I take to be Hick's subsequent suggestion, which is only to use Turing-complete languages when mathematically modeling computer programs:

> Unfortunately, such an [ideal] analysis [which is both sound and complete] is impossible for most properties of interest, such as whether a buffer is overrun (the root issue of Heartbleed). This impossibility is a consequence of Rice's theorem, which states that proving nontrivial properties of programs in Turing-complete languages is undecidable. So we will always be stuck dealing with either unsoundness or incompleteness. [176]

I certainly agree with Hicks that *if* we use a Turing-complete language, *then* we cannot ignore an important consequence of Rice's theorem, namely that "proving nontrivial properties" of our mathematically modeled computer programs (expressed in our Turing-complete language) "is undecidable." However, engineers such as Parnas resort to multiple models to describe buffer overflows (pertaining to, say, a C computer program), including models that are described with a Turing *in*complete language. The bottom line is that engineers do not have a preference for sticking to a single modeling language, nor do they advocate a Turing-complete language *per se*.

Contrary to many (if not most) programming language specialists, engineers do not want to attach *precisely one meaning* to each computer program. An engineered system can be mathematically modeled in more than one way; each model has its pros and cons. I can model my computer with both a finite state machine and a linear bounded automaton without contradicting myself. Likewise, I can mathematically model my C computer program in multiple, complementary ways, for example with a finite state machine, with primitive recursive functions, and with general recursive functions. The richness lies in the multitude of ways in which reality can be mathematically modeled, and I hope to convey this richness in the remainder of this book.

1.3   Abusing the Halting Problem

The analysis presented so far shows that, at the very least, a categorical distinction is required between computer programs and mathematical programs. The term "mathematical program" is used from now on as an abbreviation for "a mathematical model of a computer program."

Another obvious distinction that is worth making explicit at least once is the distinction between computers (which include laptops and iPads) on the one hand and their mathematical models on the other hand. Strictly speaking, then, it is wrong to say that:

> A computer is a finite state machine.

Once again, this is like speaking about a mathematical model (the finite state machine) as if it coincides with reality (the computer). But making this observation explicit in computer science, as I am doing now and as I have done in my rejected paper entitled 'Category Mistakes in Computer Science at Large,' seems to be rather unusual.

My paper on Category Mistakes was rejected by two POPL referees for some very good reasons, albeit non-technical ones. I shall have more to say about the reviews I received in Chapter 5. For now, it is important to note that my rejected POPL paper contains quoted reviews from anonymous referees of the Communications of the ACM (CACM). With regard to these CACM review comments, here is what the first POPL reviewer had to say:

> I'm not sure why CACM reviewers would ignore the difference between real-world systems and their mathematical models. I don't actually see that mistake in the quoted reviews [in your paper].

I am afraid that the reviewers of the CACM have applied faulty reasoning, and I have illustrated precisely this in my POPL paper. I shall illustrate some of this faulty reasoning in the next section, leaving the rest for Chapter 6.

Computers vs. Mathematical Models

Strictly speaking, a computer is not a finite state machine. The former is a *concrete physical* object which can be mathematically modeled by the latter, which is an *abstract* object. Here then is the first comment that I have received from a referee of the CACM:

> My laptop is a universal Turing machine, but its tape size is of course limited by the finiteness of human resources.

If you limit the tape size of a universal Turing machine, you may end up with, say, a linear bounded automaton or even an automaton that is computationally equivalent to a finite state machine. You thus end up with another *mathematical* model of computation but *not* with a laptop (i.e., a concrete physical object). To be more precise, I stress that:

> You cannot use *human* resources to limit the size of a *mathematical* object, i.e., the tape. Note that the "tape" indeed denotes a mathematical object and not a physical object, contrary to what the word "tape" seems to suggest.

You can introduce mathematical restrictions to limit the size of a mathematical object; likewise, you can use human resources to limit the size of a concrete physical object, such as a laptop. However, once again:

> A Turing machine *is* a mathematical object, it is not a computer. This is contrary to what the word "machine" seems to suggest.

I understand the CACM reviewer's train of thought. I, too, was educated as a computer scientist and I used to speak about my mathematical model as if it coincided with reality. The right way to put it, once again, is as follows:

> Placing finite bounds on an abstract object (Turing machine) does not make it a concrete physical object (laptop). Instead,

> it results in another abstract object (e.g., a linear bounded automaton or a finite state machine) that can potentially serve as another mathematical model for the physical object at hand.

I agree that these words convey a very trivial distinction. However, missing this distinction can easily lead to faulty reasoning. Only a mathematical language can be Turing complete; it thus makes no sense to question whether your iPhone is Turing universal or not (as, for instance, did almost all my computer science students at Utrecht University in 2015). Unfortunately, statements of this kind can be found all over the place, not only in peer reviews but also in articles and in books, published by reputable publishers. I have even had discussions with colleagues who start proving on the blackboard and in the classroom that my laptop *is* a universal Turing machine after all. They really think they are giving a *mathematical* proof about my laptop. I emphasize, once again, that:

> It is a mathematical model of a laptop that may or may not be Turing universal, not the laptop itself. Yet, anonymous referees of computer science's flagship journal, the Communications of the ACM, disagree with this statement and erroneously place both objects in the same category. This is where a seemingly innocent category mistake occurs.

Comparing a laptop with a Turing machine is only warranted with the proviso that we all agree we are reasoning across separate categories.

A Big Category Mistake

Grasping the significance of seemingly obvious categorical distinctions is not easy; here is yet another response that I have received from a referee of the CACM and which I have also reported in my rejected POPL paper:

> What does the undecidability proof of the halting problem for computer programs actually tell us? Like diagonalization proofs in general it may be viewed finitely as saying that, if there is a bound M on the size of accessible computer memory, or on the size of computer programs, or any other resource, then no computer program subject to the same resource bounds can solve the problem for all such computer programs.

The previous remark and the follow-up remark, presented below, are only correct if we accept the following two assumptions (both of which are wrong):

1. A computer program is a synonym for a mathematical program.

2. The mathematical program (mentioned in the previous sentence) must be equivalent to a Turing machine program and not to, say, a primitive recursive function.

The reason why the second assumption has to hold is merely because the referee is referring to the halting problem of Turing machines. Continuing with the remarks made by the anonymous referee:

> If computer program A solves correctly all halting problems for computer programs respecting bound M, then the counterexample computer program T must exceed that bound, which is why A fails for T. To solve problems of computer programs, one needs an ideal program.

This quote hints at a distinction that must be made between finite and infinite objects (with the latter being labeled "ideal"); however, the categorical distinction between computer programs and mathematical programs goes completely unnoticed. This is where a big category mistake occurs. The undecidability proof of the halting problem concerns *mathematical* programs only and not *computer* programs. The diagonal argument can only be applied to mathematical objects, not engineered artefacts. Thus, while the referee thinks this is a mathematical argument, in fact this is faulty reasoning. The referee is *not* proving something about *computer* programs but something about a *particular* mathematical model of a computer program! So much for mathematical rigor.[10]

1.4 Pluralism to the Rescue

Based on the above analysis it is now possible to provocatively define both a computer scientist of today and one of tomorrow. A computer scientist of today is somebody who conflates:

- a Turing machine and a computer,
- a Turing tape and a physical tape,

- a mathematically modeled bug and a bug encountered during program execution, and

- a mathematical object and a computer program.

Moreover, the mathematical object denoted in the last item must be a Turing machine or some object computationally equivalent to it. (I realize that programming language experts do not work with Turing machines *per se*; however, this is rather beside the point.)

The computer scientist of tomorrow, by contrast, is sensitive to the aforementioned categorical distinctions and, furthermore, is receptive to the *multitude* of answers to the seemingly simple question:

What is a computer program?

Today, we know that the "computer program" concept has acquired at least four meanings during the course of history. It can refer to:

1. a physical object à la Maurice Wilkes in 1950 and Dave Parnas in 2012,

2. a mathematical object of finite capacity à la Edsger Dijkstra in 1973,

3. a mathematical (Turing-machine) object of infinite size à la Christopher Strachey in 1973, and

4. a model of the real world that is not a logico-mathematical construction à la Peter Naur in 1985 and Michael Jackson today.[11]

Moreover, wearing my philosophical hat, I will follow Raymond Turner's recent analysis [325] and view a computer program as a technical artefact.[12] I shall define and use this fifth interpretation of what a computer program entails in Chapter 6.

The multitude of interpretations of what a computer program entails is an example of epistemic pluralism. In addition, computer scientists have not consistently followed a single interpretation of a computer program in their writings. Marvin Minsky's 1967 book, *Computation: Finite and Infinite Machines* [245], for example, uses the word "program" on page 25 to refer to data and instructions that fit in the real, finite memory of a physical computer (that is, as a physical object). On page 153, by contrast, the very same word refers to a mathematical object of "unlimited storage capacity," akin to a Turing machine. Likewise, Tony Hoare consistently used the word "computer" in 1969 to refer to a real physical device, while in his 1972 paper 'Incomputability' [181] the very same word sometimes refers to a *finite*, physical device and sometimes to a mathematical abstraction in which "*infinite* computations" can arise.

In sum, historical actors have not always explicitly distinguished between real artefacts and their models, let alone between all the aforementioned meanings of a "computer program." In the words of Bernadette Bensaude-Vincent, "epistemic pluralism is a major feature of emerging fields" [332] and I hope my readers will come to appreciate that computer science is still too young a field to be any different.

1.5 Arthur Fleck's Reflections & Raymond Boute's Scrutiny

The two cherries on the cake are Chapter 4 and Chapter 7, written by Arthur Fleck and Raymond Boute, respectively. When a software scholar receives personal reflections and detailed scrutiny from two historical actors, along with the implicit question about whether and where their narratives can be published, the scholar experiences one of his finest days.

Arthur Fleck is former chairman of the Computer Science Department of the University of Iowa. His personal and technical recollections on programming language history — on FORTRAN, ALGOL 60, EULER, APL, SNOBOL4, Smalltalk-80, FP, Miranda, and Prolog — are an absolute delight to read and I am confident that fellow historians will buy this book for Chapter 4 alone.

Raymond Boute is professor emeritus in formal methods from INTEC Department of Information Technology, Ghent University. He questions the understanding of basic concepts and the *function* concept in particular in Chapter 7. Complementary to my focus on the informal aspects of computer science, Boute advocates elementary use of symbolism to support the textual definitions at hand. I believe next generations of formal methodists can benefit greatly from reading Chapter 7.

2. TATTY TURING TALE: Turing's True Legacy Does Not Lie in Computer Building But in Programming Languages

This chapter is an improved and extended version of the present author's 'Towards a Historical Notion of 'Turing — the Father of Computer Science',' published in Taylor & Francis's *History and Philosophy of Logic (2015) 36 (3): 205-228*. Copyright remains with the present author.[13]

In the popular imagination, the relevance of Turing's theoretical ideas to people producing actual machines was significant and appreciated by everybody involved in computing from the moment he published his 1936 paper 'On Computable Numbers.' Some careful historians are aware that this popular conception is deeply misleading. We know from previous work by Martin Campbell-Kelly, William Aspray, Atsushi Akera, Allan Olley, Mark Priestley, Edgar Daylight, Pierre Mounier-Kuhn, Thomas Haigh, Maarten Bullynck, Liesbeth De Mol, and others that several computing pioneers, including Howard Aiken, J. Presper Eckert, John Mauchly, and Konrad Zuse, did not depend on (let alone were they aware of) Turing's 1936 universal-machine concept. Furthermore, it is not clear whether any substance in John von Neumann's celebrated 1945 'First Draft Report on the EDVAC' is influenced in any identifiable way by Turing's work. This raises the questions:

1. When does Turing enter the field?

2. Why did the Association for Computing Machinery (ACM) honor Turing by associating his name to ACM's most prestigious award, the Turing Award?

Previous authors have been rather vague about these questions, suggesting some date between 1950 and the early 1960s as the point at which Turing is retroactively integrated into the foundations of computing and associating him in some way with the movement to develop something that people call computer science. In this chapter, based on detailed examination of hitherto overlooked primary sources, attempts are made to reconstruct networks of scholars and ideas prevalent to the 1950s, and to identify a specific group of ACM actors interested in theorizing about computations in computers and attracted to the idea of language as a frame in which to understand computation. By going back to Turing's 1936 paper and, more importantly, to re-cast versions of Turing's work published during the 1950s (Paul Rosenbloom, Stephen Kleene, Andrey Markov Jr.), I identify the factors that make this group of scholars particularly interested in Turing's work and provided the original vector by which Turing became to be appreciated in retrospect as the father of computer science.

2.1   Introduction

It was in August 1965 that Anthony Oettinger and the rest of ACM's Program Committee met and proposed that an annual "National Lecture be called the Allen [sic] M. Turing Lecture" [1, p.5].[14] The decision was also made that the ACM should have an awards program. The ACM Awards Committee was formed in November 1965 [2]. After having collected information on the award procedures "in other professional societies," Lewis Clapp — chairman of the ACM Awards Committee — wrote in August 1966 that

> [a]n awards program [...] would be a fitting activity for the Association as it enhances its own image as a professional society. [...] [I]t would serve to accentuate new software techniques and theoretical contributions. [...] The award itself might be named after one of the early great luminaries in the field (for example, "The Von Neuman [sic] Award" or "The Turing Award," etc.) [3].

The mathematician Alan J. Perlis officially became ACM's first A.M. Turing Lecturer and Turing Awardee in 1966. Besides having been the first editor-in-chief of the *Communications of the ACM,* Perlis had earned his stripes in the field of programming languages during the 1950s and had been President of the ACM in the early 1960s. Perlis was thus a well-established and influential computer scientist by the mid-1960s. In retrospect, decorating Perlis was only to be expected. But why did the ACM honor Turing? Turing was not well known in computing at large in the 1960s and early 1970s [98]. Apparently, his name was preferred over John von Neumann's and Emil Post's, yet all three researchers had deceased by 1957 and all three were highly respected by some very influential actors in the ACM, including John W. Carr III, Saul Gorn, Perlis, and Oettinger.[15] And why are we celebrating Turing today? The latter question was posed repeatedly during Turing's centennial in 2012.

The ACM, founded in 1947, was an institutional response to the advent of automatic digital computers in American society. Only some of those computers, most of which were still under construction, were based on the principle of a large store containing both numbers and instructions — a principle that is also known today as the "stored program" principle.[16] That principle, in turn, paved the way for computer-programming advancements which were sought by men like Carr, Gorn, and Perlis. The Cold War was also responsible for massively funded research in machine translation, later also called automatic language translation. The young student Oettinger at Harvard University was one of several embarking on Russian-to-English translation projects. Another example is Andrew Booth, a British computer builder who had met in 1946 with the prominent American scientist Warren Weaver to discuss the possibility of mechanically translating one language into another. In between 1946 and 1947, Booth visited several computer laboratories in the United States, including von Neumann's lab at the Institute for Advanced Study in Princeton. By 1952, he was both an accomplished computer builder and a programmer of a mechanical dictionary. As the postwar years progressed, several computer specialists turned into applied linguists and many professional linguists became programmers.[17] Documenting this technological convergence leads to a better understanding of when, how, and why Turing assumed the mantle of "the father of computer science."

In this chapter, I describe the convergence by zooming in on the work of Booth, Carr, Gorn, and Oettinger.[18] I will show that, not until the 1950s, did computer programmers like Booth and Gorn begin to reinterpret the electronic computer in terms of the universal Turing machine. They did this with the purpose of developing higher level programming languages.[19] Turing thus assumed the mantle of "the father of computer science" for reasons that are orthogonal to the commonly held belief that he played an influential role in the design or construction of "universal computers." My historical account is primarily about the 1950s and ends with a brief discussion of Turing's 100th birthday in 2012 and with published work related to this chapter.

The take-away message in general terms is that the 1950s constitute a decade of cross fertilization between linguistics, computer programming, and logic. That decade is preferably *not* viewed as a smooth road from modern logic to computing; if there was any road at all in the history of computing, then it was most definitely from practice to theory.

2.2   Machine Translation

"See what you can do with your Russian," the student Oettinger was told around 1949 by the American computer pioneer Aiken at Harvard after the latter had corresponded with Weaver on the vexing topic of machine translation.[20] Weaver had directed American war work of hundreds of mathematicians in operations research. Fully aware of the developments in electronic computing machines, he had come to believe around 1946 that code-breaking technology from the war could help detect certain invariant properties that were common to all languages. Weaver had expressed his ambitious ideas in writing in his 1949 memorandum *Translation* [230, Ch.1] which, in turn, sparked intense American interest, including Aiken's interest, in the possibility of automatic language translation. In comparison with Booth's mechanical dictionary from 1952, Weaver's original idea on machine translation was more ambitious, namely to go "deeply into the structure of languages as to come down to the level where they exhibit common traits." Instead of trying to directly translate Chinese to Arabic or Russian to Portuguese, Weaver supported the idea of an indirect route: translate from the source language into an "as yet undiscovered universal language" and then translate from that language into the target language [230, p.2,3,15,23].

In the academic year 1949–1950 Oettinger started thinking about mechanizing a Russian-to-English dictionary. He also stayed at Maurice Wilkes's computing laboratory in Cambridge for a year where he met Turing on a regular basis. Wilkes, in turn, also visited leading figures in computing. He regularly traveled from England to the United States where he met with Aiken, von Neumann, and many others. Apart from being an accomplished computer designer, Wilkes was also practicing and advancing the art of computer programming. In this regard, around 1952, he met Perlis and Carr who were working with Project Whirlwind at MIT [342, p.31]. Perlis had obtained his Ph.D. in mathematics from MIT in 1950. Carr had done the same in 1951 and had also spent some weeks in Wilkes's computer laboratory in England together with Oettinger. The world of computer practitioners was thus a very small one: many practitioners knew each other on a personal level and several became involved in the ACM.[21]

Initially, most linguists were rather pessimistic about Weaver's memorandum, relegating his aspirations about machine translation to the realm of the impossible. Gradually, they started to see opportunities [230, p.4,137]. By 1951 the computer had made a clear mark on linguists. The Israeli linguist Yehoshua Bar-Hillel conveyed that message by making an analogy with chemistry. Chemists, he said, need "special books *instructing* students *how* to proceed in a fixed *sequential* order [. . . in their] attempted analysis of a given mixture" [26, p.158–159, my emphasis]. Likewise, special books will have to be written for the linguist, books that contain "sequential instructions for linguistic analysis, i.e., an *operational syntax*" [26, p.158–159, original emphasis]. According to the historian Janet Martin-Nielsen, American linguistics at large transformed from elicitation, recording and description before the war to theory and abstract reasoning after the war. It "rose to prominence as a strategic and independent professional discipline" [236].

Researchers knew that literal translations would yield low quality machine translation. Therefore, some of them sought methods to construct "learning organs;" that is, machines that "learn" which translation to prefer in a given context [26, p.154]. As Weaver had already put it in 1949: the "alogical elements" in natural language, such as "intuitive sense of style" and "emotional content," rendered literal translation infeasible [230, p.22].

Weaver's remarks can, *in retrospect*, be viewed as part of a grander intellectual debate in which fundamental questions were posed such as whether machines can think (cf. Edmund Berkeley's *Giant Brains, or Machines That Think* [31]). Weaver had addressed this issue optimistically in 1949. A year later, Turing's 1950 article 'Computing Machinery and Intelligence' was published [322]. It was followed up by Wilkes's 'Can Machines Think?' [341].

Weaver's memorandum was based on an appreciation for the theoretical 1943 work of Warren McCulloch & Walter Pitts, entitled 'A Logical Calculus of the Ideas Immanent in Nervous Activity' [240]. McCulloch and Pitts had essentially tried to find a mathematical model for the brain. Weaver described their main theorem as a "more general basis" for believing that language translation was indeed mechanizable by means of a "robot (or computer)."[22] In other words, according to Weaver there was no theoretical obstacle to machine translation: learning organs could, at least in principle, resolve the translation problem.

After leaving Aiken's lab at Harvard to temporarily join Wilkes's research team in Cambridge, Wilkes made clear to Oettinger that he had "no use for language translation." Instead, he urged Oettinger to address the question whether computers might be able to learn [268, p.208]. It is in this setting that, for Oettinger, "Alan Turing and others became familiars at meetings of the Ratio Club" [268, p.208]. Oettinger's work on the EDSAC led up to his 1952 paper 'Programming a Digital Computer to *Learn*' [264, my emphasis], in which he targeted an audience of psychologists, neuro–physiologists, and everyone concerned with non-numerical applications of digital computers. He connected the McCulloch & Pitts's paper to automatic digital computers and wrote that "digital computers can be made to serve as models in the study of the functions and of the structures of animal nervous systems" [264, p.1243].[23] Due to Turing's direct influence, Oettinger used the words "universal digital computer" in his paper [264, p.1244]. But, although Oettinger included both Turing's 1936 and 1950 papers in his bibliography, it was Turing's 1950 "imitation game" that mattered most to him. Oettinger was, after all, concerned with programming a real computer, the EDSAC, so that it could "learn."

After returning to Harvard, Aiken disapproved of his student's turn towards learning organs and forced him to return to his automatic language translation project [268], which would lead up to his 1954 Ph.D. dissertation, *A Study for the Design of an Automatic Dictionary*. In the immediately following years, he dived into the needs of the milk and banking industries [4, p.6]. Oettinger's papers on data processing [152, 265] and his comprehensive 1960 book *Automatic Language Translation* [266] — which was partially about programming the Univac I for the purpose of machine translation — did not contain any references to Turing and the like.

By 1961, Oettinger *was* referring — through Rosenbloom's 1950 book *The Elements of Mathematical Logic* — to the works of Alonzo Church, Turing, and especially Post [294, Ch.IV]. As professor of Mathematical Linguistics, Oettinger was giving the bigger picture of the *converging* developments that had taken place during the 1950s. "Syntactic analysis," he said, had received considerable attention, not only from the "mathematical linguists" (such as Noam Chomsky), but also from "applied mathematicians" (such as Carr and Perlis) and "mathematical logicians." The mathematical linguists were seeking algorithms for automatic translation among *natural* languages. The applied mathematicians (i.e., the computer programmers) were concerned with the design and translation of languages suitable for *programming* machines. The mathematical logicians, in turn, had been exploring the structure of formal *artificial* languages [267, p.104].

By 1963, the theoretical work of Post, Turing and several others had become common currency among some influential academics in both mathematical linguistics and in computing, particularly in automata theory and automatic programming. The professional linguist Bar-Hillel had already expressed his appreciation for "recursive function theory, Post canonical systems, and the like" in 1960 [26, p.84].[24] And in the Classification System of the December 1963 issue of the Computing Reviews, "Turing Machines" was explicitly mentioned next to "Automata."

2.3 Automatic Programming

Besides the machines housed at MIT, Harvard University, and Cambridge University, also the ENIAC, EDVAC, and IAS machines should not go unmentioned. The ENIAC and EDVAC were constructed at the Moore School of Engineering, Philadelphia, and were later housed in Maryland at Aberdeen Proving Ground, where von Neumann was a consultant for several years [11][55, p.886]. During the second half of the 1940s, von Neumann and his team built the IAS computer at the Institute for Advanced Study, Princeton [117].

During the early 1950s, Andrew Booth was, together with his wife Kathleen, writing a book entitled *Automatic Digital Calculators* (cf [41]). In the preface of their first edition, published in 1953, the authors explicitly acknowledged "their indebtedness to John von Neumann and his staff" at Princeton for "the stimulating period spent as the guests of their Project" [41, p.vii]. Reflecting on the history of computer building, the authors referred to several people, including Gottfried Leibniz, Blaise Pascal, Babbage, Joseph Jacquard, and Herman Hollerith. No mention was made of Turing throughout the whole book, except for a brief reference to the ACE computer which had been "under the direction of John Womersley, Turing and Francis Colebrook" [41, p.16].[25]

Moreover, in the section "The universal machine," the authors referred to "the Analytical Engine" of Babbage and essentially described him as the father of the universal computer.[26] In this connection, they mentioned the machines built by Aiken (1937–1944) and at Bell labs (1938–1940, 1944), describing them as "universal" and "general purpose." More specifically, the authors distinguished between a special purpose and a general purpose machine in the following manner. A "special purpose machine" was constructed to perform one set of operations, to solve one particular problem. For example, a computer that can only compute Income Tax (for different sets of input) was special purpose. General purpose computers, by contrast, were "capable of being set up to solve any problem amenable to treatment by the rules of arithmetic" [41, p.1]. The authors also clarified their notion of "universality" by noting that multiplication and addition can be defined in terms of subtraction; therefore:

> Whatever type of computing machine is projected, so long as it is to have the attribute of *universality*, it must necessarily have in its structure some component capable of performing at least the most elementary operation of arithmetic — subtraction. [41, p.22, my emphasis]

It was due to the "expense" and "difficulty" of incorporating *several* electrical or electronic components that "most modern general purpose computers" did not include square root units, dividers, and — in some cases — even multipliers [41, p.3].[27]

Some of the first computers, like the ENIAC, were — at least initially — provided with a plugging system "so that the various units [could] be connected together and sequenced to suit the particular problem to be solved" [41, p.14]. Some of the later machines of the 1940s and early 1950s, such as the EDSAC and the EDVAC, were based on the principle of a large store containing both numbers and instructions. The authors established an epistemological relationship between Babbage's work and the principle of a large store.[28] Furthermore, from the authors' perspective, the incentive to use a large store was primarily an engineering one.

> When ENIAC was under construction, von Neumann and his co-workers devoted considerable attention to the problem of the *optimum* form of a calculating machine. As a result of this it was shown that a calculating machine should have a storage capacity of the order of 1,000 [8,000][29] words and that, *given this*, both operating instructions and pure numbers could be stored in the same unit. [41, p.15, my emphasis]

Choosing to store both "operating instructions" and "pure numbers" in the same store — as opposed to storing each in a separate store — was based on practical concerns, not theoretical reasoning.[30] Likewise, the incentive among many computer builders to list all instructions consecutively in one part of the store and the pure numbers in another part (of the same store) was, according to the Booths, based on the practical concerns of "simple process control" and economic use of "memory capacity" [41, p.15, 23-24].[31]

Common storage of numbers and instructions inside the computer opened the door for new programming techniques. And, the realization that many computations can be reduced to iterative processes made it "most desirable" that data could be *erased* in *any* given memory location and be replaced by new material [41, p.23, my emphasis] — a topic which I shall return to shortly.[32]

Common storage was, however, not a prerequisite to embark on programming. Some people, such as Zuse, stored commands apart from numbers (cf. [158, p.76]). Similarly, Haskell Curry made a sharp conceptual distinction between commands and pure numbers [248]. Another important name in this regard is Aiken, as the following words from Grace Hopper in 1978 indicate:

> Aiken was totally correct in saying that the programs and the data should be in separate memories, and we're just discovering that all over again in order to protect information. [335, p.21]

Indeed, in practice today, most software never treats data as code [119].[33]

Space Cadets

Specifying the orders to accomplish a computation was a tedious task. Relieving the programmer from that task was the job of the "space cadets" — a name that Carr gave in 1953 to those programmers who wanted to invest time and money in building interpreters and compilers [342, p.210–211]. Carr and his space cadets (including Booth, Gorn, Perlis, and Wilkes) advocated automatic programming. They wanted to design "pseudo codes" — "each order of which will, in general, represent a number of orders in the true machine code" [41, p.212].

In my eyes, Carr stood out in the 1954 'Symposium on Automatic Programming for Digital Computers' [315] in that he consistently and repeatedly described *both* the true machine code and the pseudo code as a "language."[34] Carr viewed automatic programming as the problem of closing the "gap" between the "external decimal language" and the "intermediate binary language," of recognizing a "dual language system" and the need to program the "translation between the two languages" [50, p.85].[35] Finally, the "lack of compatibility" between existing digital computers was the incentive to search for a "universal language" — an idea that was mostly, although not solely, Gorn's [50, p.89][261].

Indeed, it was Carr's later good friend, Saul Gorn, who — in the eyes of Hopper — stole the show at the 1954 symposium on automatic programming. "Looking forward to what Dr. Gorn will tell us," Hopper expressed her curiosity in her opening address towards the development of a "universal code" that can be taught to mathematicians and that *each* computer installation can provide by means of an interpreter or compiler [315, p.4–5]. To counter the proliferation of "specialized codes," Gorn stressed in his talk the need for "a code more or less independent of the machine," so that the "artists, scientists, and professions can then return to a common language and creative thinking."[36]

During the first half of the 1950s, Gorn worked at Aberdeen Proving Ground where three "general purpose machines" were in use simultaneously: the ENIAC, EDVAC, and ORDVAC. This diversity in machinery prompted a "universal coding" experiment. And so Gorn developed a "general purpose common pseudo-code" that could serve as input to both the EDVAC and the ORDVAC [161, p.6].

Loop Control

At the symposium, Gorn enumerated four machine requirements that had to be met in order to provide for a universal code, and he did this by referring to Carr's prior research on loop controlled machines — research that I have yet to come across in primary sources. Gorn's account, reproduced below, serves to back up my conjecture that up and till around 1954 Carr and Gorn did *not* view loop controlled computers as practical realizations of universal Turing machines.

The first machine requirement to obtain loop control, Gorn said, was due to the universal code being "translatable by the machine into the machine language." Therefore, the machine had to be able to "construct and adjust its own commands."[37] To do the latter, "[w]e will therefore assume that commands are stored internally just like other words, so that the only way in which a word is recognized as a command is that it reaches the control circuits through the sequencing of previous commands. (In the terminology of J.W. Carr III, we are requiring a Type II Variable Instruction Machine)."[38] In addition to common storage, the machine should also have "the other two properties described by Carr as necessary for 'loop control,' namely that they be able to transfer control out of sequences, and that they be able to make decisions." Finally, as a fourth requirement, the machine should be capable of carrying out elementary commands, such as "adding, multiplying, subtracting, extracting a digit, taking absolute values, taking negatives, shifting digits, and, most elementary of all, stopping, reading, writing, and generally transferring information from one place to another."[39] To do this, the machine should have "one or more forms of *erasable* internal storage" [158, p.76–77, my emphasis].

In later years, machines were simply said to have "loop control" if they possessed three properties: "common storage, erasable storage, and discrimination" [160, p.254].[40] These properties allowed recursive problems to be programmed iteratively instead of being programmed as straight lines of code. For example, instead of coding the summation $S = \sum_{i=1}^{n} x_i$ as a straight line of code — as in

$$s_1 = x_1$$
$$s_2 = s_1 + x_2$$
$$s_3 = s_2 + x_3$$
$$\ldots$$
$$s_{n-1} = s_{n-2} + x_{n-1}$$
$$s_n = s_{n-1} + x_n$$
$$S = s_n$$

— loop control allowed the summation S to be programmed iteratively on the basis of the following three equations

$$s_1 = x_1,$$
$$s_{i+1} - s_i \mid x_{i+1} \; (i = 1, 2, \ldots, n - 1),$$
$$S = s_n.$$

The sequence of values $i = 1, 2, \ldots, n - 1$ indicates that $s_{i+1} = s_i + x_{i+1}$ had to be performed for each value of i from 1 through $n - 1$. The crux was that only one computer instruction was stored to correspond to $s_{i+1} = s_i + x_{i+1}$. The machine *modified* the addition instruction to perform the addition over and over again the required number of times. The machine furthermore decided when it had performed the addition the proper number of times [163, p.2-46]. In other words, the programmer did not have to specify up front the specific number of times the addition had to be carried through; instead, the "thinking" machine — using Edmund Berkeley's terminology [32, p.5] — could determine the specific number autonomously.[41]

Kleene's *Introduction to Metamathematics*

Practically oriented as they were, Carr and Gorn were also theoretically inclined, trying hard to make connections between their loop controlled machines and mathematical logic. One book that clearly had an impact on their thinking was Kleene's *Introduction to Metamathematics* [202]. According to Carr, automatic programmers had to deal with "the generation of systems rather than the systems themselves," and with "the 'generation' of algorithms by other algorithms," and hence with concepts akin to metamathematics [50, p.89]. Gorn, in turn, described the machine requirement of common storage by saying that "the language of the machine includes its own syntax, so that not only may the machine be directed to tell us something, but it may also be directed to tell us something about the way in which it tells us something." For example, the machine could, besides generating the output of a calculation, *also* print out "the storage locations chosen for the key commands and variables," and, by doing so, tell the programmer something about the way it computes (i.e., the way in which it tells something). Gorn linked this to the "quasi-paradoxical properties, as revealed in the researches of Gödel" [158, p.75-76].

Gorn continued appropriating ideas from metamathematics in a 1955 technical report, which he distributed to several colleagues, including Carr, Perlis, Mario Juncosa, David Young, Herman Goldstine, and the logician Barkley Rosser [159]. On the one hand, Gorn remarked, every computational procedure by an automatic machine "must follow the steps of a constructive proof."[42] On the other hand, Gorn argued, "every constructive existence proof, with a few slight modifications, provide[s] a possible method of solution. (What we have just said is an intuitive way of stating 'Turing's Thesis'. See reference [202])."[43]

Later on in his report, Gorn described an "ideal general purpose machine" as one that is

> effectively equivalent to a 'universal Turing machine.' (See reference [320]. The universal Turing machine can copy the description of any special purpose Turing machine and imitate its operation by means of these copied specifications.)

By referring to a universal Turing machine, Gorn made clear that, to him in 1955, an "ideal" machine had unlimited storage capacity. Existing machines were, however, not ideal in that they could not store exact real numbers but only finite approximations of them. This deficiency, Gorn noted, could, without additional precaution on behalf of the programmer, easily result in computing errors.[44]

To summarize, by 1955, Carr and Gorn *were*, at times, viewing a universal Turing machine as a conceptual abstraction of a loop controlled computer.

"General-Purpose Machines"

In a 1956 booklet *The electronic brain and what it can do* [162], Gorn distinguished between "older machines" in which "switches were set externally" and manually, and the newer "four-address machines" such as the EDVAC in which "each order contains a part that automatically sets the switches for the carrying of the next order." The bottom line was never having to plug or unplug a single wire ever again. Because orders were coded as numbers, these newer machines gained "a flexibility and power that were undreamed of only a few short years ago" [162, p.42, 58].

A year later, Gorn wrote a letter to his good friend Perlis who, as editor-in-chief of the *Communications of the ACM*, published the letter on the first page of the very first issue of the magazine [272]. In that letter, Gorn continued reflecting on EDVAC's design which, he said, had "introduced the major step forward" in computer history "of having a common variable storage of instructions and data" [272, p.2].[45] The "manipulation of hardware" had been replaced by the "programming of coded symbols."

In his letter to Perlis, Gorn also recalled how "it has been recognized that all general purpose machines, from Edvac on, are essentially equivalent, any one being capable of the same end results as any other, with varying degrees of efficiency."[46] This recognition did not happen overnight, nor did it become common knowledge to all researchers in automatic programming.[47] It was few men like Carr and Gorn who repeatedly, during the 1950s, tried to see the metamathematical — and in their eyes, grander — picture of their practical accomplishments.

In the same spirit, Gorn noted that two fundamental principles had been recognized in recent years.[48] First, that hardware and programming were within-limits interchangeable. Second, that "our general purpose machines" are equivalent with "a certain, as yet ill-defined, universal command language." During the late 1950s, Gorn, Carr, and Perlis played leading roles in the USA in the development of the universal programming language ALGOL 60 (cf. [259]).[49]

Gorn and Carr were engaged with Burks's research agenda on cellular automata, and with the overarching theme of *thinking* and *learning* machines. The "mathematical machine of the future," Carr wrote in 1959, "may well be an *even more general* purpose machine" than what we have today [61, p.12, my emphasis]. Carr's dream was to achieve "Completely Automatic Programming." This meant that, once a proper algorithm was developed, it should be possible for the machine to decide on its own machine code. The machine of the future "should be allowed to engineer its own construction," in contrast to present "fixed code machines" that are "not self-encoding" [61, p.11].

Emil Post

Researchers in the Soviet Union were also interested in the generation of algorithms by other algorithms; that is, in the design of "compiler-producing compilers." Versed in Russian, and having visited several computer laboratories in the Soviet Union, Carr was aware of the Russian state of the art. Particularly impressed by Markov Jr.'s 1954 book *Theory of Algorithms* [197], Carr described Markov's "language" as something that "could be used directly as an input to compilers which are to perform symbol manipulation (such as compiler-producing compilers)." Moreover, Carr mentioned that several Russian papers [186, 200, 287] discussed the "extension of such language application to [...] other indirect reasoning processes [such] as theorem proving, *natural language translation*, and complex decision making" [61, p.190, my emphasis].

Responsive to the books of Rosenbloom and Markov (and to the Russian literature in general), Carr became attracted to the strong underlying kinship between Perlis's work on automatic programming and Chomsky's research in mathematical linguistics [65–67]. The kinship was due in no small part to Post whose work, as Carr observed, had influenced Perlis via Markov and Chomsky via Rosenbloom.[50]

The fields of automatic programming and machine translation were *converging*. Perlis's "string language manipulator," for example, was proposed to be used not only with "symbolic operational programs," but also with "algebraic language compiler-translators" and "natural language translators" [61, p.230]. Likewise, Carr noted, the construction of automatic dictionaries had "already begun" with "algebraic languages as Fortran, Unicode, and GP at one end, and with the language translation experiment at the other [end]" [61, p.259–260].

Alan Turing

In 1959, Carr used the notion of a "Universal Turing Machine" to explain a fundamental insight about automatic programming: one can "simulate a new not-yet produced digital computer" on an older computer, provided that there is sufficient storage for the simulation. The first such situation, Carr reflected, "was the preparation of an interpretative program simulating the IBM 704 on the IBM 701 before the former was completed" [61, p.236, 239].

Much in line with Carr and Gorn's 1954 reception of metamathematics and self-referential arguments in particular, Carr elaborated on the idea of having a computer simulate *itself*.

> If one universal machine can simulate any other machine of a somewhat smaller storage capacity (which is what Turing's statement on universal machines means), it should therefore be possible for a computer to simulate a version of itself with a smaller amount of storage.

The practical implication of self simulation, Carr continued, is that it allows programs to be run that "would perform in the standard fashion, and at the same time print out pertinent information such as location, instruction, previous contents of [the accumulator], and the previous contents of the address of the instruction" [61, p.240].

Insights into interpreters, compilers, and — what we today call — program portability was, in short, what a universal Turing machine had to offer to Carr and some of his space cadets.[51]

Also British space cadets, like Booth and Stanley Gill, used Turing's 1936 paper to paint the bigger picture of what they had been accomplishing independently of Turing, and what they were going to accomplish with Turing's theory as a road map. Booth accredited Turing as the founder of automatic programming in his opening address at the Working Conference on Automatic Programming of Digital Computers, held at Brighton in 1959.[52] It was Turing, Booth proclaimed, who "first enunciated the fundamental theorem upon which all studies of automatic programming are based" [157, p.1]. Booth continued as follows:

> In its original form the theorem was so buried in a mass of mathematical logic that most readers would find it impossible to see the wood for the trees. Simply enunciated, however, it states that any computing machine which has

> the minimum proper number of instructions can simulate
> any other computing machine, however large the instruction
> repertoire of the latter. All forms of automatic programming
> are merely embodiments of this rather simple theorem and,
> although from time to time we may be in some doubt as to
> how FORTRAN, for example, differs from MATHMATIC or the
> Ferranti AUTOCODE from FLOW-MATIC, it will perhaps make
> things rather easier to bear in mind that they are simple
> consequences of Turing's theorem. [157, p.1]

Booth was appropriating Turing's work in the field of automatic
programming, and he most likely did this on the basis of Carr and
Gorn's earlier metamathematical insights.[53]

Equally noteworthy, for the purpose of documenting Turing's scholarly
legacy, is Booth's historiographical remark about Turing's 1936 paper.

> Why was it, then, that Turing's original work, finished in 1937
> before any computing machine of modern type was available,
> assumed importance only some years after machines were
> in common use? The reasons, I think, stem entirely from the
> historical development of the subject. [157, p.2]

According to Booth, the first computing machines were used almost
exclusively by their constructors and, hence, by people who were
intimately aware of their internal construction. It took some years before
the machines were used for scientific applications, devised by people
who were and wanted to remain ignorant of the machine itself and,
hence, had to rely on automatic programming techniques [98, p.24–25].

Two Metaphors

Apart from Booth's opening address, also Gill discussed some practical
implications of Turing's work at the 1959 workshop in Brighton in his talk
"The *Philosophy* of Programming" [my emphasis]. On the one hand, Gill
noted, automatic programming can be viewed as a "language translation
problem;" that is, "from the language used by the programmer into the
actual instructions required by the computer." This language metaphor
— still perceived as novel by many in 1954 — was a well established
metaphor in 1959.[54] On the other hand, Gill continued, "*another* way of
looking at automatic coding is based on the idea expressed by *Turing*;"
namely that "one computer can be made to imitate or to simulate another

computer" [157, p.182, my emphasis]. Turing's work, Gill continued, implies that

> an actual computer, together with a suitably designed program, [is] equivalent to another computer which may not actually exist as a separate entity. When looked at in this way, an automatic coding scheme behaves rather like the skin of an onion, which if removed, reveals another onion underneath. [157, p.182]

Gill articulated this *onion-skin metaphor* in terms of the words "hierarchies of coding schemes" and "hierarchy of programming languages." To explain this hierarchy, Gill made an analogy with a hierarchy of natural languages. An English man may, at a later stage of his life, supplement his native language with a glossary of terms related to his profession, say, nuclear physics. Then, when studying a particular topic in nuclear physics, he might further supplement his language with a list of particular symbols and definitions, and so on. Likewise, for a computer which is supplied with coding scheme No. 1, one can supplement it in turn with scheme 1a or scheme 1b, "each adapting the system in a little more detail to some particular class of problems" [157, p.186].[55]

By 1960, Gorn was preaching about *languages* and Turing *machines* all over the place. He called his research agenda "Mechanical Languages and their Translators" and he described his agenda in terms of the words: "decidable languages," "universal mechanical languages," and "universal formal mixed languages" [327, p.29].[56] Most striking (to the historian) is his *summarized* curriculum of Electrical Engineering (1959–1960) in which he used the words "Turing machine" three times(!) and, specifically, as follows:

- Digital Computers - Engineering Logic

    - Elements of number theory; Turing machines and the foundations of computation theory; [...]

- Theory of Automata

    - Advanced problems in varieties of automata, including combinatorial nets, sequential nets, and Turing machines. Languages for describing the behavior of automata. A brief account of recursive function theory, formal axiomatic systems and Goedel's theorem; the relation of these to automata and their significance for an understanding of the general nature of machine computation.

- Introduction to Digital Computers: Systems and Devices

    - Principles of mechanization of computations derived from special and universal Turing machines. General purpose computers: system organization, input-output, logic, storage, and memory devices and timing. [. . .] [327, p.33–38]

Clearly, to Gorn and Carr, Turing machines did not solely belong to the domain of automata theory. Turing machines were also relevant in the practical context of automatic programming. And, as we have seen, other automatic programmers, like Booth and Gill, were also delivering speeches about some *practical* implications of Turing's *theoretical* work, during the late 1950s.[57]

2.4 Half a Century Later

Like Booth and Gorn in the 1950s, computer scientists in the 21st century have the tendency to construct a set of "foundations" for computer science in which ideas and events that had not been understood as connected when they happened are retrospectively integrated. (This, I believe, is characteristic of scientific progress!) Subsequent generations of computer scientists then tend to assume that these things have always been related. The connection between universal Turing machines and computers is a major example. On the other hand, critical, self-reflecting, questions *are* put into the open every now and then. For example, the question "Why are we celebrating Turing?" was asked at several Turing events in the year 2012, the centennial of Alan M. Turing. Nobody gave a historically-accurate answer however, hence my incentive to write the present chapter.

From a sociological perspective, I postulate that a majority of researchers gathered together in 2012 to celebrate their common belief that theory really does come before practice: mathematics and logic are indispensable in computing; that is, after all, what von Neumann and Turing have shown us, is it not? Or, as Robinson puts it:

> We can today see that von Neumann and Turing were right in following the logical principle that precise engineering details are relatively unimportant in the essential problems of computer design and programming methodology. [292, p.12]

Many researchers, including myself, concur with the idea that one should first grasp the logical principles (of, say, a software project) before working out the "engineering" details. But I object when people, and even the community at large, rephrase history in support of their cause. Robinson, for example, *also* writes that "Turing's 1936 idea had started others thinking" and that "[b]y 1945 there were several people planning to build a universal [Turing] machine." [292, p.10]. But, in fact, most people during the 1940s did not know what a universal Turing machine was, nor had they come across Turing's 1936 paper or a recast version of his work.

Another example of myth creation is due to S. Barry Cooper. He writes that

> The universal [Turing] machine stored programs and so gave us an understanding of the modern stored-program computer before anyone had even built a real one. [75, p.777, original emphasis]

Thus, according to Cooper, theoretical insights led to practical achievements. I have given ample evidence to suggest that he's got the story backwards. The theoretical insights came after years of practical work, and, moreover, by theoretically inclined programmers who were few in number.

> [continued:] This caused all sorts of problems. [...] It was easy to get the mathematical overview. The problem was to match it up with reality. And this was a problem with a practical aspect. Even a toy avatar of the abstract machine was hard to make. The engineers eventually came up with clever solutions to the problem: the EDVAC in Pennsylvania, the Manchester 'Baby', Maurice Wilkes's EDSAC, and the Pilot ACE growing out of Turing's own attempts to build a computer at the National Physical Laboratory (NPL). [75, p.777]

Here we have a beautiful example of how myths are created and sustained. Cooper, a main organizer of several Turing events in 2012, is painting a romantic, yet historically inaccurate, picture of Turing's role in the history of computing. But, if *you* disagree with the previous passage and, hence, do not share the common prevailing belief that Turing is the inventor of the computer and that all engineers were depending on him, then *you* are in for some peer pressure.

> [continued:] But to this day, there are engineers who find it hard to excuse (or even understand) Turing's reputation as the 'inventor' of the computer. [75, p.777]

Alas, also trained historians have fallen into the trap of merely asserting, rather than proving, Turing's influence on computer building. Simon Lavington, for example, writes in his recent book *Alan Turing and his Contemporaries: Building the world's first computers*, that

> *Turning theory into practice* proved tricky, but by 1948 five UK research groups had begun to build *practical stored-program* computers. [220, p.xiii, my emphasis]

> All of the designers of early computers were entering unknown territory. They were struggling to build practical devices based on a novel abstract principle — a *universal computing machine*. It is no wonder that different groups came up with machines of different shapes and sizes, [...] [220, p.8, original emphasis]

These words contradict what ironically seems to be Lavington's main thesis: that Turing had almost no direct influence in computer building [220, Ch.8]. A similar, yet more severe, critique holds for Mahoney's entire oeuvre — *Histories of Computing* [234] — which I have presented elsewhere [99] and partly in Chapter 1.

As the previous examples illustrate, getting Turing's legacy right isn't easy. By hardly mentioning Turing in their joint book, *Computer: A History of the Information Machine* [57], Campbell-Kelly and Aspray put Turing into the right context. Similar praise holds for Akera's prize-winning book *Calculating a Natural World: Scientists, Engineers, and Computers During the Rise of U.S. Cold War Research* [5]. (A critical side remark here is that Akera's book is also about computing in general and, as the present chapter shows, Turing's work did play an important role in this more general setting.) The research of Liesbeth De Mol [246], Olley [269], Priestley [280, 281], Daylight [98, 100], and the recent work by Haigh, Priestley, and Crispin Rope all attempt to

1. clarify what we today call the "stored program" concept, and/or

2. put Turing's and von Neumann's roles into context by explaining what these men *did* do or how their work *did* influence other computing pioneers.

Finally, although Mahoney uncritically documented Turing's legacy [99], he also presented an impressive coverage of the rise of "theoretical computer science" — a topic that, to the best of my knowledge, has so far only been scrutinized by Mounier-Kuhn [251–254]. Two findings of Mounier-Kuhn that complement this chapter are, in brief, that

1. Turing's work, and modern logic in general, only gained importance in French computing during the 1960s, and

2. it is a "founding myth of theoretical computer science" to believe that the "Turing machine" was "a decisive source of inspiration for electronic computer designers" [253].

2.5 Closing Remarks

The existing literature is rather vague about how, why, and when Turing assumed the mantle of "the father of computer science." I have partially filled that gap by showing that Turing's 1936 paper became increasingly relevant to the influential ACM actors Carr and Gorn around 1955. Perlis's reception of Turing has yet to be documented in future work.

Carr and Gorn were inspired by re-cast versions of Church, Post, and Turing's original writings. These versions included:

• Rosenbloom's 1950 *Elements of Mathematical Logic* [294],

• Kleene's 1952 *Introduction to Metamathematics* [202], and

• Markov Jr.'s 1954 *Theory of Algorithms* [197].

In the late 1950s, also British automatic programmers, including Booth and Gill, appropriated ideas from Turing's 1936 paper.

All aforementioned men were attracted to the "universal Turing machine" concept because it allowed them to express the fundamental interchangeability of hardware and language implementations (of almost all computer capabilities). Unsurprisingly, insights such as these came during a decade of cross-fertilization between logic, linguistics, and programming technology — a decade in which computing was still a long way from establishing itself as an academically respectable field.

Just like some space cadets, also the reader may have noticed a strong similarity between Weaver's 1949 "interlanguage" or "universal language" on the one hand, and the notion of an intermediate machine-independent programming language on the other hand.[58] Another example is the pushdown store which, as Oettinger explained in his 1961 paper, served a unifying technological role across the domains of automatic programming and machine translation [267].

Understanding the birth of *computer science,* a term which I use in this chapter for the first time, amounts to grasping and documenting the aforementioned technological convergence. On the theoretical side of this convergence, Church's, Turing's, and especially Post's work became increasingly relevant — a topic that concerns the notion of undecidability (see [98, Ch.2]). On the practical side, the universal Turing machine played a clarifying role in that it helped some experienced programmers — i.e., the aforementioned space cadets — to grasp the bigger picture of what they had been accomplishing in conformance with the language metaphor, a metaphor that became well established during the 1950s. To be more precise, the universal Turing machine paved the way for the complementary onion-skin metaphor, thereby allowing the space cadets to view their translation problem (from one *language* into another) in the interrelated, operational, setting of *machines*.[59] In modern terminology: the programming *languages* ALGOL 60 and FORTRAN can be viewed as equivalent computational tools to that of a universal Turing *machine.* From this theoretical perspective, it does not matter which programming language or which computer one prefers, they are all equivalent and (to a very large extent) interchangeable.

By the end of the 1950s, two metaphorical seeds had thus been sown for future advances in computer programming. The concepts of language and machine became increasingly interchangeable. Dijkstra, for example, would in later years frequently describe a layer in his hierarchical design as either a language or, equivalently, as a machine.[60] The origins of Structured Programming are, in my opinion, firmly rooted in the language metaphor and the onion-skin metaphor [81, Ch.1].

Viewing the universal Turing machine as a mathematical model of a general-purpose computer — as Booth, Carr, Gill, and Gorn did in the second half of the 1950s — also meant that the *ideal* general-purpose computer was one of *infinite* storage capacity. Nofre et al. appropriately use the words "conceptualization of the computer as an *infinitely* protean machine" [261, my emphasis] (to describe this transformational point of view) but they do this without mentioning the pivotal role the universal Turing machine played in this regard.[61] Moreover, many historical actors never joined Gorn et al. in viewing computers as infinitely protean machines. Dijkstra, Parnas, and others frequently insisted throughout their careers not to ignore the finiteness of real machinery, even when reasoning abstractly about software.[62]

It was *some* mathematicians — most notably Carr, Gorn, and Perlis — who in the field of automatic programming tried to seek unifying principles and who advocated making a science. In 1958, for example, Carr explained his desire for "the creation of translators, techniques for using them, and, finally, a *theory* of such formal translators. [...] The development of 'automatic problem solutions' requires formalism, interchangeability of procedures, and computability of languages if it is to become a true discipline in the *scientific* sense." [60, p.2, my emphasis].

Carr, Gorn, and Perlis played a large part in helping form and shape the ACM during the 1950s and 1960s. Under the initiative of Oettinger, the ACM chose to honor Turing in 1965. This was well *before* Turing's secret war work in England came into the open. In other words, Turing's influence was already felt in automatic programming, automata theory, and other research fields before popular books were published about his allegedly significant role in the makings of the first universal computers.

It was Babbage who, rightly or wrongly, was put front and center during the 1950s as the father of the universal computer (cf. Alt [10, p.8]). During the 1950s and 1960s, Turing was never portrayed as the father of the universal computer. Since the 1970s, popular claims have been made, literally stating that Turing is the "inventor of the computer" or "the inventor of the universal computer" — see the romantic accounts of Copeland [76], Davis [87, 89], Dyson [117], Leavitt [221, 222], and Robinson [292] for some typical examples; see Burks's fitting rebuttals [54, 55] and van Rijsbergen's and Vardi's sober reflections on Turing's legacy [289, 329].

"What would Turing be doing today if he were still alive?" — a question that was posed multiple times in celebration of Turing's Centennial. He would be countering the now-prevailing belief that he is "the inventor of the computer"!

3. TINY TURING TALE: Turing's 1936 Paper Did Influence the Dutch Computer Builder Willem van der Poel

This chapter is a translated and modified version of part of the present author's 2016 contribution in *De Geest van de computer: Een geschiedenis van software in Nederland*, published by Uitgeverij Matrijs, Utrecht.

Several historiographies of computer programming cover American and British developments; relatively few narratives that are written in English discuss the progress of continental European countries, such as the Netherlands. One goal of the present chapter, then, is to integrate parts of the history of Dutch computing in an international setting. Specifically, I will outline advancements made in the Dutch cities Delft, The Hague, and Amsterdam during the 1950s and embed these in an American-British context. Moreover, I will compare and contrast the technical achievements of Willem van der Poel in Delft and later in The Hague with those of Gerrit Blaauw, Edsger Dijkstra, and Adriaan van Wijngaarden in Amsterdam, and link the accomplishments of these four actors to those of American and British researchers, particularly in connection with the advent of machine-independent programming and the reception of Alan Turing's work.

The question that lies at the heart of the present chapter is the following:

Did Turing's 1936 paper 'On Computable Numbers' influence the early history of computer building?

"Yes, certainly" and "no, definitely not" are often-heard answers. A third, more nuanced, response acknowledges a diversity of local computing habits in the 1940s-1950s. Some historical actors became acquainted with Turing's 1936 paper early on, while many others did not. Some researchers depended directly or indirectly on its contents, while many others accomplished great feats even without knowing who Turing was. The successful Dutch computer builder Willem van der Poel is one of those few men who not only read about, but also applied, Turing's 1936 universal-machine concept in the history of early post-war computers. Van der Poel thought and programmed very much like Turing himself and was industrially much more successful than Turing. Van der Poel's story is told here in English for the first time.

Introduction

The Netherlands entered a period of reconstruction and austerity after World War II. Nevertheless, its government did invest a considerable amount of money in research. In Amsterdam, for instance, the government helped found a Mathematical Center in 1946, which from 1947 onwards contained a computing department. Adriaan van Wijngaarden was appointed the first head of that department and led a small research team with the main purpose of building and subsequently using automatic calculating machinery.

The Dutch government also supported the technical school in Delft (later called the University of Delft) by investing money in the construction of the ARCO, a calculating machine that provided researchers with the opportunity to advance the state of the art in optics. Willem van der Poel, a young physics student in Delft, developed the basic ideas for the ARCO before moving to the PTT, the Dutch government's post and telephone company in The Hague, where he designed and built a variety of increasingly powerful machines.[63]

The Dutch technology company Philips, by contrast, did not play a very active role in the computing industry during the 1950s. Its physics laboratory, Nat Lab, did build the PETER between 1953 and 1956, a calculating device that would later aid researchers in the fields of crystallography, switching functions, and cyclotron design. Philips also constructed both the PASCAL and the STEVIN, both between 1956 and 1960, but refrained from building commercial machines due to an agreement with IBM. Philips had promised IBM not to enter the computer market; IBM, in turn, had guaranteed to continue buying its computer components from Philips rather than manufacture them itself.[64]

The years of post-war reconstruction came with a strong tendency to mathematically model problems from the real world, that is, to express physical phenomena in the language of mathematics. In this setting, Dutch mathematicians eagerly made themselves available for social service. On the one hand, many of them became opposed to an ivory tower mentality, that is, criteria related to technology and commerce would from now on determine whether a mathematical solution was adequate or not. On the other hand, however, focusing on the engineering problem *per se* was only part of the mathematician's task; revealing the underlying mathematical structure of the problem was to be at least as important [6, Ch.8].

The Dutch centers that fostered the construction of new calculating machinery also became the places where one could ponder the use of such machines. For the cases in which researchers used a program — that is, a sequence of instructions — to control their machine, the machine was perceived as much more flexible than the common, mechanical, calculating machines of the past. Reflecting on the use of such machinery resulted in new ideas on how to program. The research centers in Delft and The Hague, on the one hand, and Amsterdam, on the other, developed their own approaches to computer construction and computer programming. In this chapter I shall contrast both approaches by discussing some of the writings of four historical actors: van der Poel, Blaauw, Dijkstra, and van Wijngaarden.[65]

In the following sections I convey and refine some key findings put forth in the writings of Gerard Alberts & Huub de Beer [7], Adrienne van den Bogaard [39], and Edgar Daylight [33] — findings that have not yet been brought to the attention of non-Dutch-speaking scholars. My specific contribution is that I contextualize van der Poel's early reception of Turing's 1936 notion of universality; that is, I contrast his reception of Turing's writings with those of John Carr and Saul Gorn around 1955 (see Chapter 2). The Americans Carr and Gorn differed from van der Poel in that they abstracted away the finiteness of their computing machinery and, more precisely, treated a universal Turing machine as an ideal model of a general-purpose computer. Furthermore, and unlike the three aforementioned actors, the Amsterdam researchers Blaauw, Dijkstra, and van Wijngaarden were hardly, if at all, influenced by Turing's work during the 1950s (see [98]). I will argue that, unlike van der Poel, the Amsterdamers were just as eager as Carr and Gorn to abstract away from their machine. However, while Carr and Gorn had access to operational machinery, the Amsterdamers abstracted away from non-existent machinery, i.e., from machinery that was still under construction.[66]

In alignment with the writings of Dutch historians, my findings support the emerging thesis that the absence of operational computing machinery in Amsterdam during the 1950s turned out to be a blessing in disguise in later years, both nationally and on an international level.

3.1   Delft

Around 1947, Nicolaas G. de Bruijn — who had just been appointed professor and would later become the inventor of AUTOMATH — submitted a project proposal to the Hogeschoolfonds of Delft. His objective was to obtain funding for the construction of an "all-round" calculating machine, a "self-thinking" device. The project was approved and a grant was secured. In de Bruijn's words:

> The objective of this research proposal is to prepare the construction of a modern all-round calculating machine which will aid the T.H. [Technische Hogeschool of Delft] and surrounding organizations. [...] Such a "self-thinking" device can be constructed entirely from logical elements such as, for example, the normal telephone relays [...]. Alternative, yet equivalent, approaches can be based on mechanical, pneumatic, and electronic relays. [51, my translation]

The goal of de Bruijn's project was to build a calculating machine that would carry out computations in optics. An important part of the research was the selection of adequate materials for the machine, which could be mechanical, pneumatic, electrical, or electronic.

Leen Kosten, head of the Mathematics Department of the Central Laboratory of the PTT in The Hague ensured that de Bruijn's project in Delft would receive multiple relays "on loan." Kosten's contribution allowed for the construction of an electrical machine, the ARCO. Based on a selector switch, the ARCO could be configured to use one of several "recipes," i.e., one of several hard-wired programs.

The actual project work was assigned to van der Poel, a technical physics student who had a clear and realistic picture of the limitations of calculating machinery. In 1947, he wrote:

> Tasks such as indefinite integration and reduction of an equation to its most simple form cannot, of course, be carried out by the [ARCO] machine since they rely on subjective criteria. We have yet to reach the stage in which machines will think for us. [274, p.1, my paraphrased translation]

Immediately after graduating in 1950, van der Poel joined Kosten's team, leaving other physics students to finalize the ARCO's construction, a machine that was later known as "the TESTUDO" (turtle) due to its extremely slow execution speed.

Van der Poel's cautious attitude in 1947 with regard to computing machinery was more than merely an example of Dutch austerity; it was also a result of his comprehensive knowledge of the computing literature. Besides referring to published writings in connection with electronic circuits and the construction of calculating machines, van der Poel pointed to the recent reports of Burks, Goldstine, and von Neumann. Even the mathematical logic of Hilbert and Ackermann and Turing's 1936 article 'On Computable Numbers' formed part of his knowledge base.[67] Van der Poel had characterized the most essential problems for the Delft project as "programming problems" [274, my translation].

> Solving a problem by programming amounts to choosing an appropriate set of instructions. Of course, the challenge is to accomplish as much as is feasible with as few instructions as possible. [274, p.60, my translation]

The electrotechnical engineer Kosten, in turn, had obtained his Ph.D. during World War II on the design of analog machines with the purpose of simulating congestion problems in telephone traffic. With van der Poel now on his side, the choice was quickly made to build a radically new digital calculating machine, the PTERA, which differed greatly from the TESTUDO. Many years later, van der Poel would describe the transition from the TESTUDO to the PTERA as a transformation from "pre-von-Neumann" machines to "post-von-Neumann" computers.[68]

> [The TESTUDO] was still a pre-von-Neumann machine. At that time I did not yet have the idea of storing instructions and numbers in the same memory. I became aware of that idea when I read the famous report 'Preliminary discussion of the logical design of an electronic computing instrument' of Burks, Goldstine and von Neumann [...] [278, p.8, my translation]

According to van der Poel's recollections, it was the writings of von Neumann et al. which had made him realize that a program could be stored *inside* the machine. During his 1988 retirement speech, van der Poel recalled how he had obtained von Neumann's report from the Mathematical Center of Amsterdam [278, p.8].

The writings of von Neumann et al. also made a difference on the international scene. However, contrary to popular belief, and as alluded to in the previous chapter, many if not most computer designers who considered using a small instruction set did *not* do this in order to practically realize a universal Turing machine (see also Haigh et al.'s recent book *ENIAC in Action* [170].) An exception in this regard other than Turing himself was the successful Dutch computer builder van der Poel, as we shall see later.

3.2   Amsterdam

Delft was not the only Dutch city that was eagerly trying to enter the computer age. Van Wijngaarden's team in Amsterdam was anxiously building their first machine, the ARRA, although this machine never operated properly and, with hindsight, its persistent malfunctioning resembled Amsterdam's incompetence in the new field of computer building. It was in the political arena that the ARRA played an important role. The politically minded van Wijngaarden succeeded in using the ARRA in 1952 to secure extra research funding from the government. In combination with the arrival of the experienced Dutch computer builder Gerrit Blaauw from the USA, the Amsterdam team set out to make a difference on a national and later on an international scene.[69]

Born in 1924, Blaauw was an electrical engineer from Delft who had spent one year in Pennsylvania (but *not* at the Moore school), followed by four years at Harvard (1949–1952) under the mentorship of Howard Aiken at the Harvard Computation Laboratory. Blaauw contributed to making the MARK III computer operational and wrote his Ph.D. dissertation on the design of the MARK IV. He left Harvard for Amsterdam in 1952, carrying a suitcase filled with computer equipment, and joined van Wijngaarden's team in September of that same year [37, 95].

Blaauw could not have wished for a better American education, for by 1944 Aiken had already succeeded in building an operational and reliable automatic calculating machine. Moreover, in contrast to many contemporaries, Aiken presumed and anticipated that multiple components of his machines would malfunction during operation;

failure and recovery were the norm for Aiken. Therefore he used older and more reliable technology for the construction of his machines and ensured that his components were interchangeable.[70]

One anecdote goes that it only took some 20 minutes for Blaauw to convince van Wijngaarden's ARRA builders, Bram Loopstra and Carel Scholten, that he had a lot of expertise to offer. Blaauw's first Amsterdam machine was finalized in December 1953. Although it was also called ARRA, it had nothing else in common with the first malfunctioning ARRA. Amsterdam's second ARRA machine and the later-built FERTA (from 1955) were almost identical, and both were of Blaauw's making.[71]

Before Blaauw's arrival, Amsterdam had had little or no expertise in computer building. It was primarily Blaauw's design methodology, carrying Aiken's signature, which enabled van Wijngaarden's team to build operational computing machinery. Some of Blaauw's accomplishments were his introduction of a clocked machine, uniform building blocks, pluggable and therefore exchangeable parts, a neat design method, and structured documentation.

Blaauw never really felt at home in Amsterdam, despite his good partnership with Dijkstra, and he continued his career in the USA in 1955. His departure for IBM left Amsterdam behind with a reputable computer center in Western Europe.[72]

3.3 Edsger Dijkstra

Edsger Dijkstra joined van Wijngaarden's computing department in March 1952 on a part-time basis, spending most days of the week in Leiden studying theoretical physics. In 1953, when the second ARRA was still under construction, Dijkstra wrote its functional specifications, along with the excerpts presented below. Unlike Blaauw, Loopstra, and Scholten, Dijkstra was not an engineer; he was not in a position to grasp all the physical intricacies of a computing machine. For each computer that was being built by van Wijngaarden's team during the 1950s, Dijkstra would treat its design as a contract between himself (the programmer) and the engineers Blaauw, Loopstra, and Scholten.[73]

Specifically, Blaauw, Loopstra, and Scholten focused on *how* the machine would operate, while Dijkstra worked on *what* the machine would be able to do, as described by Dijkstra himself in 1953:

[T]his machine will be described as far as is relevant for

the person who <u>uses</u> the machine: we shall describe <u>what</u> the machine does, and not <u>how</u> the machine works. [106, Dijkstra's emphasis, my translation]

Dijkstra also made extensive use of analogies in his research. For example, in order to transition from *how* a computer's memory works to *what* it does, he resorted to analogies of marbles in a bag and a teacher holding a piece of chalk and a sponge. In his words:

> Implementing a memory (which stores numbers) like marbles in a bag would be impractical. Instead, the memory is analogous to numbered cells written on a blackboard. With chalk in one hand and a sponge in the other hand, the sponge is used only when the contents of a cell have to be changed. Erasing with the sponge happens automatically; that is, 'forgetting' is not a distinct operation. [106, p.1, my paraphrased translation]

In retrospect, then, Dijkstra extracted the essentials of his programming task by discarding several physical details of the computing department's machine. Moreover, but as was also common in numerical analysis at the time, Dijkstra was reasoning in 1953 *from* a general mathematical problem *to* a computer program. In his words, the process was:

"1. mathematically formulate the problem

2. mathematically solve the problem

3. select/construct the numerical process (in accordance with 2.)

4. programming: detailed construction of the processes in 3. ...

5. coding: write out the program in machine code ..." [106, my translation]

Furthermore, Dijkstra emphasized that the programming in 4. was applicable to *several* machines, not only to the ARRA. Coding in 5., in turn, *was* machine specific. Or, to put it in the words of the historian van den Bogaard, by 1953 Dijkstra was making a clear distinction between programming and coding [39, p.133].

To recap, then, on the one hand Dijkstra was already reasoning abstractly to a noticeable extent during his first years as a programmer; this will become more apparent when contrasting Dijkstra's writings with van der Poel's work in Section 3.4. On the other hand, I hasten to remark that Dijkstra's abstractions were not as extensive as in later years, nor were they unique at the time.

In his 1953 report, Dijkstra was still making trade-offs, much in line with the work of his fellow engineers. For instance, he enumerated six conflicting ideals of the programmer:

"1. maximum speed, to execute the program

2. minimum memory, to store the program

3. maximum safety

4. maximum accuracy

5. maximum flexibility

6. a maximum bird's eye view (of the program)" [106, my translation]

Dijkstra immediately noted that "The first two ideals are already to some extent conflicting ideals." To demonstrate this conflict between the first two ideals, Dijkstra elaborated on the difference between a "Stretched program" (which also went by the name of a straight-line program in the 1950s) and a "Cycle program." To create the former, a programmer would have to write each operation repeatedly; that is, as many times as was needed. To create the latter, the programmer had to write the operations once and place them in a subroutine which, in turn, would determine during program execution how many times to carry out the cycle.[74]

Dijkstra subsequently noted that more memory was needed to implement the "Stretched program" than the "Cycle program," but that the "Stretched program" executed faster than the "Cycle program." This typical engineering trade-off had machine-specific implications, as Dijkstra noted. When working with a slow machine with a large memory, the programmer would opt for the "Stretched program." When dealing with a fast machine with a small memory, the programmer would choose the "Cycle program" instead. This machine-specific reasoning (specialization) stands in sharp contrast to Dijkstra and van Wijngaarden's later work on ALGOL (generalization) in the late 1950s and 1960s, as described in my book *The Dawn of Software Engineering: from Turing to Dijkstra* [98].

3.4   Willem van der Poel

While Blaauw's arrival in the fall of 1952 allowed the Amsterdam team to fulfil a large part of their computer building aspirations, van der Poel was already acquiring international fame in The Hague with his 1952 paper 'A Simple Electronic Digital Computer' [275]. In this paper van der Poel demonstrated his esthetically-driven style of computer building. The topic of the paper was the design of the ZERO machine which he later described as his "most beautiful machine ever." While awaiting the delivery of more computer equipment for the construction of the PTERA, van der Poel temporarily constructed his ZERO machine in 1952.[75]

ZERO's beauty exemplified economy of design and logical minimalism, as the following examples show. Van der Poel used the same register to serve both as accumulator and control register. He avoided expensive multiplication and division components in hardware by programming them with addition. He implemented the addition of two numbers within one and the same electronic component by means of bit-wise addition sequentialized in time (these last two design choices led to slow computers). Finally, van der Poel resorted to four "functionally independent bits" [275]. One bit b_1 expressed whether the machine's instruction had to read something from ($b_1 = 0$) or write something to ($b_1 = 1$) the drum. Another bit b_2, independently expressed whether the accumulator had to be cleared ($b_2 = 0$) or not ($b_2 = 1$). The two bits together ($b_1 b_2$) then defined four possible combinations: 00, 01, 10, and 11. Since the value of the first bit did not depend on that of the second and vice versa, no control box was required and hence less equipment was needed, which resulted in small and cheap computing machinery. Led by esthetics, van der Poel was willing to pay the price that "not all combinations were practically useful;" indeed, only four of the seven instructions in his ZERO machine made practical sense.[76]

The ZERO only existed for a couple of months and was quickly dismantled in favor of the PTERA. The latter became operational in September 1953 and would serve several departments of the PTT for a total of 12 years. However, the PTERA was also a slow and cheap machine, as reported in the newspaper 'het Nieuw Utrechts Dagblad' on September 15, 1953 in an article entitled 'Brilliant Work Delivered by PTT-Engineers':

> [The PTERA] is smaller than contemporary machines built in England, the United States and some other countries. Due to its compactness, the machine is slower but also much less expensive. [my translation]

Van der Poel's next project was the design of an even more powerful computer, the ZEBRA, which he described at length in his 1956 Ph.D. dissertation [276]. Van der Poel re-used ideas from the ZERO on his ZEBRA computer, in which 15 functional bits were contained in a 33-bit instruction word. Following similar principles of esthetics elsewhere in his design, he strove for "complete duality," realizing full well that "this is seldom of practical importance."[77]

Neither the PTT nor Philips showed interest in building the ZEBRA. Although Konrad Zuse's firm, ZUSE, did, this was not for long. It was the English company STANTEC which eventually manufactured the ZEBRA from 1957 onwards, delivering dozens of ZEBRAs throughout Europe. The lively correspondence between several ZEBRA users across Europe led to the formation of the ZEBRA club with van der Poel as a well-respected guru.

We will see that it was van der Poel's computer building style and his complementary approach to computer programming which made him an influential computer pioneer in the 1950s.[78]

Simplicity

Van der Poel's designs of "simple" and "very simple" machines, as reflected in the names of his machines, were strongly influenced by a theoretical sense of beauty. During the 1950s, only few researchers connected mathematical logic with the construction of a computer; van der Poel was one of these. He applied Turing's 1936 notion of "universality" in the design of his ZERO computer. Specifically, and as illustrated in van der Poel's 1952 article 'A Simple Electronic Digital Computer,' he strove for the simplest computer design that remained "universal" in the exact mathematical sense of Turing [275, p.376].

Van der Poel's strong interest in mathematical logic becomes even more apparent in his 1956 Ph.D. dissertation 'The *Logical* Principles of Some *Simple* Computers' [276, my emphasis]. He noted that in propositional logic it is possible to rewrite implication ($\rightarrow$) in terms of disjunction ($\vee$) and negation ($\neg$). For example, the implication $a \vee b \rightarrow c$ can be rewritten as $\neg(a \vee b) \vee c$. In other words, implication ($\rightarrow$) is logically superfluous.

It should be remarked at this point that similar observations with regard to logical minimalism were to some extent also made by Andrew Booth and other contemporaries. Booth knew that the multiplication of two numbers can be reprogrammed in terms of addition, thereby making a multiplier a redundant component of a machine. Booth pursued such simplifications in order to obtain smaller computer designs and cheaper constructions.

However, van der Poel went a few steps further than Booth et al. He also observed that in propositional logic one can rewrite *all* connectives, including implication ($\rightarrow$), disjunction ($\vee$), and negation ($\neg$), in terms of just one connective: the Sheffer stroke ($\uparrow$).[79] Furthermore, and most importantly, van der Poel referred to this logical ideal in his Ph.D. dissertation and subsequently explained at length and by analogy how, in principle, he could build a *universal* computer — in Turing's sense of the word, not in Booth's sense — that only contained one machine instruction. This theoretical limit, he said, was something he had almost reached with his ZERO machine which had consisted of only four machine instructions [276, p.105].

On the one hand, considerations from mathematical logic thus clearly played a direct role in van der Poel's computer-building work. On the other hand, it should also be remarked that van der Poel appropriated Turing's 1936 theory solely in the setting of *finite* machinery. According to van der Poel, the universal Turing machine was *not* an ideal model of a modern computer. He did not want to abstract away the finiteness of real machines. In his words:

> In "On computable numbers", TURING [41] does not exercise restraint as regards the size of the machine or the extent of the store. Therefore the class of the computable numbers is infinite. From a practical point of view it is, however, better to restrict oneself to finite machines. Then the latter can no longer be called universal in the sense of TURING, because they cannot generate all "computable numbers". [276, p.102]

Furthermore, van der Poel noted in his dissertation that undecidable problems, as pondered by Turing himself in 1936, become trivially solvable in the context of finite machinery; that is, in the context of finite-state machines.

Contrast this latter observation, then, with John Carr and Saul Gorn's reception of the mathematical logic literature. By 1955, both Americans were appropriating the universal Turing machine as a model of a general-purpose computer and, by doing so, were conceptualizing the computer as an infinitely protean machine, as recounted in Chapter 2.

Abstracting away from the real, finite, world was a prerequisite for entering the abstract realm of mathematical logic. Doing so allowed

Gorn to paint a grand picture of computing by the end of the 1950s.[80] He noted that two fundamental principles had been recognized in recent years: first, that hardware and programming were, within limits, interchangeable, and second that "our general purpose machines" are equivalent to "a certain, as yet ill-defined, universal command language." During the late 1950s, Gorn, Carr, and their American colleague, Alan Perlis, played leading roles in the USA in the development of the universal programming language ALGOL 60 (cf. Nofre [259]).

Optimum Coding and Underwater Programming

Led by economy of design and logical minimalism, van der Poel built intrinsically slow computers. To resolve this problem, he exploited his thorough understanding of the machine's construction *while* programming. Specifically, he perfected two already existing techniques, optimum coding and underwater programming, and thereby demonstrated a strong similarity with Turing's own machine-specific programming habits (see, e.g., Campbell-Kelly [56]).

Optimum coding essentially meant accessing the drum economically, e.g., by interleaving instructions and data on the drum in accordance with the way the program would behave. The drum was after all the slowest part of the computer. The common practice of independently storing instructions and data on the drum resulted in several drum rotations (during program execution). To reduce the number of drum rotations, van der Poel opted for a less orderly solution by interleaving the instructions and the data on the drum in accordance with the order in which they would be called by the processor [331, p.26].

Underwater programming amounted to minimizing the drum accesses, e.g., by copying an instruction I from the drum to the registers and subsequently modifying the contents of the registers in order to transform I into the next instruction I', and I' into I'', and so forth. Until the drum was accessed a second time, the program was executing "under water," using van der Poel's terminology [93]. The reduced number of accesses to the drum allowed the program to maintain a high speed of execution. It was not easy to circumvent the drum by exclusively resorting to the registers. To be successful in this regard, the underwater programmer had to have a thorough understanding of *how* the machine had been built.

The historian Eda Kranakis elaborates on the connection between van der Poel's functional bits and his underwater programming in the following way. Depending upon how the bits were set up in any given instruction, the ZEBRA could be made

> to draw a new instruction from the drum at the same time that it was performing an addition with a number drawn from a register; or it [could] be made to form a new instruction in the control unit by drawing an instruction from the drum and then modifying it by the contents of a word in the register. [212, p.76]

Underwater programming, according to Kranakis, referred to programs that were, on the surface, very short and simple. But, when run, the instructions — by way of the functional bits — would

> re-execute themselves many times over and even alter themselves in the process so that ultimately a very long and intricate computation would be carried out. The term, 'underwater programming' was intended to convey the analogy with the activity that goes on in a lake or sea, undetected from the surface. [212, p.76]

It should be noted that Kranakis's interpretation of "underwater programming" is, unlike mine, not based solely on recent interviews but also on at least one primary source [277].

Through support for optimum coding and underwater programming techniques, the ZEBRA became an economically competitive machine, as the firm STANTEC conveyed in 1958:

> Due mainly to the entirely new programming concept and the simple logical design upon which it is based, the cost of the STANTEC-ZEBRA is far less than might be expected for a machine of its capabilities.

Without this machine-specific programming technology, van der Poel's esthetically pleasing machines would have remained extremely slow and, as a result, economically unattractive. The price van der Poel paid for his esthetics was that his programs were incomprehensible to almost everyone except himself.[81]

3.5 Dijkstra versus van der Poel

In contrast to van der Poel in The Hague, van Wijngaarden and Dijkstra in Amsterdam were decoupling the activity of programming from the machine. Both men lectured on programming from 1951 onwards and their first lecture notes appeared in print in 1955 with the following title: 'Programming *for* Automatic Calculating Machines' [my translation, my emphasis]. In their lecture notes, they wrote:

> The most important thing, however, is that the program *does* express what *happens*, but *not* what all of this *entails*, . . . In short, there is a need for a comprehensive overview of programs, a notation in which — and thanks to the omission of accessories — the essentials are not submerged, and where moreover the *function* of the program parts is stipulated. [. . .] The notation of flow diagrams accommodates this. This notation does not directly rely on the machine code of the underlying machine; therefore, we are ensured of a valuable form of generality of the flow diagrams. On the other hand, the notation is less efficient when compared to tricked programs [à la van der Poel] . . . in which one has to exploit several specific — often haphazard! — characteristics of the underlying machine [339, my translation, original emphasis].

In other words, van Wijngaarden and Dijkstra wanted to program independently of the machine. Taking specific properties of the machine into account, they said, would only obscure the programmer's task. These aspirations conflicted with the "tricked programs" of van der Poel, as the historian Adrienne van den Bogaard confirms in her writings about the ARMAC machine.

Van den Bogaard contextualizes Dijkstra's 1959 Ph.D. dissertation, *Communication with an Automatic Computer* [107], which addressed the following technical problem: How can the computer execute instructions without hampering the simultaneous operation of the input and output devices? Dijkstra abstracted away from the machine and defined a general problem. In Dijkstra's words:

> We prefer to direct our attention to a general problem that arises in connection with the problem of synchronisation. [. . .] The so-called 'interruption' has been built into the machine to execute a specific communication program within a period of time, that is independent of the main program. [107, Dijkstra's emphasis]

Van Wijngaarden and Dijkstra used the word "general" extensively in the 1950s (see [39, p.138,139,141]) and later (e.g., [337, p.16]) to convey their perspective on programming. Likewise, their contemporaries Christopher Strachey and Peter Naur described the Amsterdam school with the words "general" and "very general" in 1962–63 [98, p.59,70]. An articulated account of what is meant by "general," when used to describe what is distinctive about the Amsterdam school and Dijkstra's work in particular, is presented by means of several examples in my previous writing 'Dijkstra's Rallying Cry for Generalization: the Advent of the Recursive Procedure, late 1950s — early 1960s' [98, Ch.3].

3.6 Closing Remarks

Besides van der Poel in Delft and later in The Hague, it was primarily American and British researchers who successfully designed and constructed the world's first automatic, digital, calculating machines. Some of the first computers, such as the ENIAC and van der Poel's ARCO, were — at least initially — provided with a plugging system "so that the various units [could] be connected together and sequenced to suit the particular problem to be solved" [41, p.14]. Some of the later machines of the 1940s and early 1950s, such as the EDSAC, EDVAC, and van der Poel's PTERA, were based on the principle of a large store containing both numbers and instructions, also known as the stored program principle.

Computer specialists frequently traveled across the Atlantic to meet each other in person. Maurice Wilkes from Cambridge (England), for instance, traveled to the USA to meet Aiken in Harvard and von Neumann in Princeton. Likewise, Carr visited Wilkes's laboratory which housed the EDSAC. The Dutch researchers also visited Wilkes's laboratory, including van der Poel, van Wijngaarden, and Dijkstra. The latter attended Wilkes's 1951 programming summer school and, in return, received a punched card from Wilkes containing a message for van Wijngaarden, stating that he should hire Dijkstra as soon as possible. Van Wijngaarden did so on March 1, 1952 and promptly assigned Dijkstra the task of programming the non-existent ARRA.[82]

Unlike van der Poel, men like Carr and Gorn were researching machine-independent programming techniques. Having had operable, modern, computing machinery available for several years, the reader may be tempted to believe that these experienced American programmers were therefore in pole position to pursue abstractions away from the machine in the early 1950s. In fact, Carr and a few others called themselves the "space cadets" in 1953 in order to openly distance themselves from what they considered to be the mainstream in computing research [342, p.210–211]. Van der Poel's machine-specific programming stood in sharp contrast to the space cadets' research agenda.

However, van der Poel's stance towards machine-specific programming also conflicted with that of his fellows in Amsterdam. Van Wijngaarden and Dijkstra were, just like the space cadets, intentionally abstracting away from computing machinery, albeit for a different reason: the Amsterdamers simply did not have a machine to work with until 1954 (while Gorn, for instance, already had three operable machines by 1954 [261]).

By 1957, Dijkstra was using the word "pseudo-machine" [105, p.103] and discussing how one can use a "superprogramma" (i.e., a "superprogram" in English) to communicate with the machine in ways akin to conversing with a colleague, as opposed to instructing a slave [105, p.106]. It was Amsterdam's quest for machine-independent programming which gained momentum during the second half of the 1950s, rather than van der Poel's fixation on the machine itself. Around 1960, the Amsterdamers played a pivotal role in the design and implementation of ALGOL. Two years later, van der Poel's computer-building days ended: he became professor at the Technische Hogeschool Delft and started participating in IFIP's working group on ALGOL. As professor, van der Poel also taught logic and followed up on his Ph.D. dissertation in which he had associated the Sheffer stroke with his own one-instruction Turing universal computer, a computer that existed only on paper and which served as an ideal model for his four-instruction ZERO machine [276, p.105].

The findings in the present chapter, along with those of Alberts, de Beer, and van den Bogaard in previously published Dutch texts, support the emerging thesis that the absence of operational computing machinery in Amsterdam turned out to be a blessing in disguise in later years. Many years later, Dijkstra described his 1953 distinction between *what* a computer does and *how* a computer works as a "separation of concerns;" this principle played an important role in the later careers of both Dijkstra and Blaauw [33].

4. TRUEHEARTED TALE about Turing-Award Winners: Personal Reflections on Programming Language History — by Arthur Fleck

About the Author

Arthur Fleck was born in 1936 in Chicago Illinois. He received the Bachelor's degree in Mathematics from Western Michigan University in 1959, and the Master's and Ph.D. degrees in Mathematics from Michigan State University in 1960 and 1964, respectively. After spending one year as an Assistant Professor in the Electrical and Computer Engineering Department at Michigan State, he joined the Computer Science Department at the University of Iowa. Except for a year at the University of Virginia, he remained at the University of Iowa. He served as department chairman there for two terms from 1984 to 1990, and retired in 2006.

Prelude

This chapter sets forth personal impressions about programming languages I have encountered throughout my career. I assess my experience with programming languages extending back 60 years, and

Figure 4.1: Arthur Fleck

cover both well-known and some less well-known languages. Over the years a vast number of programming languages have been conceived so that even being aware of all their names is a challenge (e.g., [340]). This writing involves only a small selective list, and even for those languages that are mentioned, my emphasis is on the context of my experience with the language and related personal digressions rather than definitive language coverage.

4.1   A First Step

The first programming language I learned was machine (and assembly) language for the IBM 650. The IBM 650 was the earliest mass-produced computer with nearly 2000 produced between 1954 and 1962. This was the computer used in my first programming course, taught by Mr. Jack Meager, at Western Michigan University in 1956. Our computer had a 1000 word rotating magnetic drum memory, each word consisting of ten decimal (technically bi-quinary) digits. A considerable complication for our class was that at the time, Western had no computer at all available for student use! Fortunately this class was small (under ten), and our enterprising instructor arranged for the class to make after-hours use of the data processing center at Whirlpool Corporation some 50 miles away. It was amazingly generous (and brave) of a major corporation to turn a group of undergraduates loose in their data center, but it was a fantastic opportunity for us. So twice per week, the class spent a long evening traveling, and preparing (on card punch equipment) and debugging programs conceived during the intervening time. Since the opportunities to "try it" were so few, this extremely limited access to a computer instilled a lasting appreciation for meticulous care during program creation. The only 'text' I recall for this course was the machine manual [187].

Machine (and assembly) language is so linguistically primitive that I question referring to it as a "language." Assembly language only provides symbolic reference to opcodes and memory addresses rather than machine numeric, and it includes almost no linguistic structure. Better might be calling these "notations" rather than languages, but that's not the usual convention so I won't protest any further.

Western Michigan University had only this one course in computing at this time. The course was offered in the Mathematics Department (it would be ten years before Computer Science Departments began to emerge at U.S. universities). I was a math major and found this to be one of the most interesting courses I took. I did repeat the course one additional time (at the instructor's invitation, but for no credit) and gained some further experience, but there were no other locally available opportunities at that time. Despite its brevity, I found this a highly informative, in fact compelling, initiation into computing.

A noteworthy coincidence occurred during my senior year at Western Michigan University. I was invited to join the honorary science society, and the induction ceremony revolved around a dinner. The after-dinner speaker at that function was Dr. Gerard Weeg, a mathematics professor and Computer Laboratory staff member at Michigan State University. He was a fantastic after-dinner speaker, brilliantly mixing humor with an intriguing discussion of computer research projects taking place in their Computer Lab. It was fortuitous that I had some background relevant to this fascinating talk since most in the audience were not that fortunate. I had no idea at that moment that either Michigan State or Dr. Weeg would play any role in my future.

4.2 Getting Serious

Shortly before completing my undergraduate degree I rather belatedly decided to pursue a graduate degree. I applied for admission to several graduate mathematics programs, including teaching assistantship applications. My admission applications were successful, but the teaching assistantship applications were not. I finished my undergraduate degree in mid-year and the middle of the academic year proved to be an unfortunate time to seek a teaching assistantship. Since we had already started a family, significant financial assistance was a necessity, and the way forward looked uncertain. However, it was my good fortune that the Head of the Mathematics Department at Western, Dr. James Powell, learned through personal contacts of the availability of a research assistantship in the Computer Laboratory at Michigan State University.

So as fate would have it, in January of 1959 I entered the graduate mathematics program at Michigan State University (MSU), and began a research assistantship in the Computer Laboratory under the direction of Prof. Weeg! The Computer Laboratory was the computing service unit of MSU. A computer named MISTIC (see Figure 4.2) had been built at the Lab under the direction of Dr. Lawrence von Tersch, and became operational in late 1957. MISTIC was a duplicate of ILLIAC I at the University of Illinois, and was the first computer on the MSU campus. MISTIC was a binary machine with 40-bit words and 1024 words of internal (vacuum tube) memory. Input-output was via paper tape and teletype equipment. Over the years of its operation, staff at the Computer Laboratory eventually constructed and added an expanded magnetic core memory, and connected punched card equipment for input-output.

In the Computer Lab at MSU I learned assembly languages for several machines. Of course, one of these was MISTIC, the sole computer at the Computer Lab for several years after my arrival. But in the early 1960s the Lab was anticipating retiring MISTIC through the acquisition of commercial computers from the Control Data Corporation (CDC). The first arrival was the CDC 160A (a small desk-size computer), followed a year or so later by the CDC 3600 (a powerful mainframe computer). These machines had architectures that contrasted sharply with that of the IBM 650 and MISTIC. The CDC 160A and CDC 3600 were "modern" computers with index registers, asynchronous I/O, interrupt driven processing, magnetic tape and high-speed peripherals. The assembly language of each of these machines bore the mark of its architectural character. My programming work during this period was largely on systems programming tasks. In fact, the CDC 160A and CDC 3600 could be cabled together to form an early commercial multi-processor system, and one of my programming projects employed that configuration. Several long-term users of the Computer Lab had substantial investment in software for MISTIC and were deeply concerned about its impending retirement. To dispel these concerns I was assigned to construct a

Figure 4.2: MISTIC computer, constructed at MSU in 1957. (Photo courtesy of Michigan State University Archives.)

simulator for MISTIC on the new computer(s). The simulator was written (in assembly language) for the CDC 3600, but the 3600 did not have a paper tape reader, and as a primary input device of MISTIC, paper tape reader operation was embedded in much of its software. However, the CDC 160A did have a paper tape reader, so the coupled computers were used to program a simulator that provided unimpaired execution of any running MISTIC program. This provided early (about 1963) experience with multi-processor programming.

The next programming language I used was FORTRAN. FORTRAN originated in 1954 with an IBM team led by John Backus. A compiler first became available for the IBM 704 in 1957. FORTRAN adopted the level of abstraction of familiar mathematical expressions rather than machine instructions. It relieved the programmer of so much tedious detail, and so greatly increased programmer productivity that its use in many applications was compelling. This advantage soon forced other computer companies to develop FORTRAN software for their own computers. However, compiler construction was ad hoc and not well

> If E and F are expressions, and F is not floating point unless E is too, and the first character of F is not $+$ or $-$, and neither E nor F is of the form $A**B$, then
> $$E**F$$
> is an expression of the same mode as E. Thus $A**(B**C)$ is an expression, but $I**(B**C)$ and $A**B**C$ are not.
> The symbol $**$ denotes exponentiation; i.e. $A**B$ means A^B.

Figure 4.3: FORTRAN syntax rule for exponentiation

understood, and programmers with any related experience were in extremely short supply. So FORTRAN did not become broadly adopted or available on other manufacturers' machines until into the early 1960s. Since the MSU Computer Lab operated CDC computers, it was the early 1960s before I encountered the language. During the intervening period the language had evolved through three versions (see Backus [17]).

FORTRAN was initially described in a careful but informal way [188], typically in the context of examples. For instance, a list of six "Formal Rules for Forming Expressions" was given, and in Figure 4.3 you see rule number 6 from that list.

One consequence of this informal description was that the most convincing argument one could make for or against the legitimacy of a FORTRAN program containing an uncertain construct was whether or not it worked in "the" compiler. But whose compiler — numerous computer manufacturers were providing a "FORTRAN" compiler, and they varied significantly in what was admissible, and in how an admissible program worked. Although FORTRAN's position was dominant in those days, this problem seriously troubled its use as it spread from IBM to industry-wide acceptance. This difficulty could make portability of a FORTRAN program between different manufacturers' computers as troublesome as writing the program from scratch all over again. This situation persisted until 1966 when the first standard [328] was established (but significant compatibility difficulties still lingered on).

In addition to my programming duties in the Lab, I had immediately joined an on-going reading group of research assistants organized by Prof. Weeg. We met regularly to read and discuss research papers of current interest such as Mealy [241], Moore [249], Ginsburg [150], and Rabin & Scott [283]. We did not read the renowned paper by Turing [320], but instead studied Martin Davis's book on computability [86]. This background provided me with crucial early preparation and direction for eventual thesis research. My thesis research developed in algebraic automata theory [127–129]. This complemented my practical experience in the Computer Lab and provided my entry into theoretical Computer Science research, and attracted the participation of a number of other researchers to this area.

After completing my Ph.D. in 1964, I joined the Electrical and Computer Engineering Department at Michigan State University with the title of Assistant Professor of Computer Science. This title was a bit remarkable in that a Computer Science department did not officially exist at MSU until five years later (and in the College of Engineering), but I did appreciate the gesture. In my first academic appointment I taught a year-long advanced undergraduate computer programming course of my own design, and continued an appointment in the Computer Lab as head of systems programming.

During that first year of teaching at MSU I learned my next two programming languages. One of these languages was ALGOL 60. ALGOL 60 [20] was developed by an international committee with the purpose of creating an international algebraic language "to describe computational processes," while remaining "as close as possible to standard mathematical notation." The committee members drew on experience with working problem-oriented languages as a basis, but they also introduced new ideas with ALGOL 60. The overarching idea of "block structure" was one innovation. In addition to providing the organizational basis for subroutines, it allowed for nesting of program elements with scoping rules that provided for localization of identifiers at any point in a program. For procedures/functions two distinct means of transmitting arguments (by name or value) provided enhanced generality. Also, there was multi-assignment, dynamic for-statements, dynamic arrays with generalized indexing, etc. Finally, there was one other important feature missing in FORTRAN that was provided and fostered in ALGOL 60 — recursion (how this came to be is an interesting story, see Daylight [94]). Following its success in ALGOL 60, recursion became a feature of nearly every programming language to follow (even, eventually, FORTRAN). The collection of features in ALGOL 60 provided new challenges for compiler writers and fostered significant research into techniques needed to accommodate them (e.g., [286]). The collection of features, their generality, and their interactions introduced various subtleties that led to the need for a clarifying revision of the defining report [21] three years later, and this is a point that will arise again later.

A crucial part of the process of describing this new programming language was to incorporate a notation to precisely detail what was admissible in the language, and do so in a manner completely independent of any specific compiler or computer. The conceptual idea for the notation was due to John Backus [19], although almost simultaneously there were several closely similar but apparently independent developments that I'll return to later. This notation came to be known as BNF, an acronym for Backus Normal Form, or sometimes alternatively Backus-Naur Form to credit the editor of the ALGOL 60 Report, Peter Naur. Although devised by John Backus, it was Peter Naur who championed its use [256] in the defining report and coalesced the description of ALGOL 60 around BNF. As one small indication of the change that the use of BNF brought to the description of syntax, consider the ALGOL 60 BNF rule (from the revised report [21] actually) for an exponential in an expression:

<factor> ::= <primary> | <factor>↑<primary>

And compare that with the earlier example of FORTRAN's syntax rule (Figure 4.3). The use of BNF for the language description made it not only more clear and succinct but at the same time provided greater generality. John Backus received the ACM Turing award in 1977 for his work on FORTRAN and BNF, and Peter Naur received the ACM Turing award in 2005 for his work on ALGOL 60.

ALGOL 60 was an influential programming language for several reasons. It incorporated programming language features that stimulated years of study and discussion (e.g., [49, 131, 205, 235]) that promoted advances in the area. Through its reliance on BNF, another aspect of ALGOL 60 was the utter clarity of its syntax and its avoidance of quirks and special cases. By this one example, BNF was instantly established as the means to describe programming language syntax. BNF is classified as a metalanguage — a language used to describe other languages. So while it does not go into a list of programming languages, BNF is a linguistic construct, and being confident with its use proved to be indispensible to subsequent learning of other programming languages. BNF has been used to describe almost every language developed after ALGOL 60, and it provides a vital tool for pursuing foundational work on programming languages. While the significance of ALGOL 60 for programming languages is profound, experience has shown that BNF has had an even greater and more long lasting impact. It is striking that at this early stage, for the second time John Backus played a critical role in programming language development.

The second programming language I learned during that first year of teaching at MSU was IPL V. Early versions of the language were developed at the RAND Corporation in the late 1950s. Allen Newell and a group at RAND and Carnegie Institute of Technology developed the

IPL V version [258] (the first public version) that was initially released in 1960 for several IBM computers and subsequently implemented on other manufacturer's machines (including CDC). While FORTRAN and ALGOL 60 were designed primarily with numeric processing in mind, IPL V was designed for symbolic (non-numeric) processing. It was intended for adaptive problem-solving tasks involving symbol manipulation such as formal theorem proving. The heart of IPL V was the linked-list data structure — this was a data abstraction that encouraged thinking at a higher level, allowed great flexibility and generality of data organization, and provided for dynamic space utilization. A prominent feature of IPL V was the inclusion of an extensive library of list manipulation operations that facilitated symbolic computations. Linguistic characteristics of IPL V were primitive, and I did not use it beyond the teaching context that first year. However, work on IPL V inspired much additional interest and its operations were to be encountered many times over in subsequent languages (e.g., [38]). During this time period, linked-list processing techniques developed rapidly by a variety of sources, both practical and formal (e.g., [238]). Years later (1975) Allen Newell was co-recipient of the ACM Turing award, in part for work on list processing.

While there would be a germination period, my early experience and particularly my programming languages exposure during my first year as a faculty member, would exert a growing influence on my research activities. My initial research interest would evolve over time toward formalisms that were to become vital foundations for programming and programming languages.

4.3 Settling into an Academic Career

In 1965 I accepted an offer to join the Computer Science Department at the University of Iowa in its inaugural year. This new department was located in the College of Liberal Arts and had a close connection with the Mathematics Department. This was a good match with my background and interests, and initially I also held joint appointments in both Mathematics and the Computer Center. Professor Weeg had left MSU before I finished my Ph.D., and joined the University of Iowa as the Director of the Computer Center with a joint appointment as a Professor of Mathematics. In a remarkably short time he accomplished the formidable task of gaining approval for a new department in a discipline that was not yet well established! He served as the first Chair of this department as well as the Director of the Computer

Center. While a variety of courses and programs could be found at numerous institutions, for department-level academic units with 'Computer Science' in their name, a list by Ralph London [231] shows only 10 departments existing in the U.S. by 1965, with the first starting in just 1962 (at Purdue University).

FORTRAN, ALGOL and assembly language, plus COBOL, dominated the programming landscape in the U.S. throughout the 1960s. At the University of Iowa (which used IBM facilities), the FORTRAN language was used in programming classes starting in the early 1960s, and it continued to be used in the first programming course until 1976. However, a number of other languages had begun to appear – e.g., APL, Basic, SNOBOL, and several variations of ALGOL. In fact, a widely referenced "Tower of Babel" cover on the Communications of the ACM in January 1961 listed over 70 languages in common use, and emphatically reflected a rising concern over the language proliferation phenomenon. In the mid-1960s in response to this programming language proliferation, IBM mounted an effort to unify the language diversity that continued to expand by developing a broad-spectrum language that incorporated features from FORTRAN, ALGOL, COBOL and other languages. The goal was to provide a single language suitable for both scientific and business applications as well as system programming. The initial design of the language was carried out by a team composed of IBM staff members plus members of the IBM user group SHARE, and was initially called NPL [284]. Due to a conflict with a prior use of this name, this language was soon renamed as PL/I.

PL/I freely adapted features from the languages indicated above and integrated them in a coherent way. But it was innovative in expanding with features not available in its predecessors. It provided three classes of storage allocation — static, automatic (stack-based), and (programmer) controlled (heap-based). It also included exception handling features and rudimentary multitasking. It included string handling, pointers and complex numbers. Extensive defaulting conventions were included and its large set of keywords were also available as program identifiers. These features greatly complicated the task for compiler writers. In his Turing Award presentation, Hoare [180] who participated in the PL/I design process, recalls his dissatisfaction with the ambition of both the design process and its results. He speaks of unsuccessfully urging the elimination of features viewed as "dangerous" and expresses the judgment that PL/I was a "technically unsound project."

IBM developed an implementation of the language for its machines (released in 1966) and the language was quite successful in that context. As the language's popularity spread, it began to be adopted for teaching by a growing number of universities. However, in those days both the typical mode of computing ("batch processing") and the size and lack of efficiency of the IBM PL/I compiler made it impractical for large classes. This led Cornell University to develop (on IBM computers) an efficient in-core compiler system PL/C [250] suited for this purpose. In 1970 the beginning programming course at the University of Iowa was expanded into a two course sequence. The first course continued to be taught with FORTRAN, but with the support of the PL/C system, the second course in the sequence used PL/I. And in 1976 the first course of our beginning sequence also began using PL/I. Development of implementations of PL/I by other computer manufacturers was problematic, and by 1980 interest in the language had declined substantially. While PL/I did not achieve its motivating purpose (language proliferation subsequently accelerated rather than declining), it had a significant impact on programming language ideas, goals and ambitions. The successful implementation of this language encouraged subsequent language designers to elevate linguistic matters and reduce the emphasis on efficiency when considering language goals. This alone has led to noteworthy contributions to the development of programming languages, of course, all made feasible by amazing increases in machine capacity.

In the latter half of the 1960s another language attracted a good deal of my attention. This was the language EULER, introduced in two papers published by Niklaus Wirth and Helmut Weber in 1966 [346]. This language was a derivative of ALGOL 60 and provided generalizations of some features, although it abandoned static typing. Its primary attraction to me was the inclusion of a complete formal (and machine independent) definition of both the syntax and semantics. BNF provided the standard means of accurately describing programming language syntax, but was not directly useful to describe semantics. Wirth & Weber developed a formal means of precisely describing semantics and used EULER as their case study demonstration. They attached the generation of code for a virtual list-oriented machine to the parsing process for a restricted type of grammar. Each production of the grammar for EULER was attached to a code fragment that described an execution effect. As the parsing process identified the occurrence of a production, it also generated the corresponding code. This research was an elaborate harbinger of Knuth's attribute grammars [206] although it avoided the construction of a derivation tree by direct augmentation of the parse procedure, and thereby could only operate with what are known as "synthesized attributes" in attribute grammars. However, this was sufficient for a comprehensive semantic description.

For several years, I used EULER and its associated formal description in a graduate course on programming language foundations. It provided a superb illustration of the complete formal description of a language. The language description included detailed operational development for parsing, code generation, virtual machine and interpreter, providing a model suitable for creating an implementation of EULER. This was done under my oversight at the University of Iowa (and by others at other universities), allowing experimentation with some relatively intricate EULER programs. The EULER description exposed the complication and subtlety that derive from ALGOL's block structure scope rules, call-by-name parameter transmission, procedures as arguments, etc. I gained much better insight into these issues by augmenting the study and use of the EULER description with that implementation effort, and I believe it was a fruitful educational tool. While there were a number of efforts at connecting formal specification of language syntax and semantics with implementation, I regarded the EULER project as a model of conciseness and completeness.

Although in the early 1970s, the first programming course at the University of Iowa still taught FORTRAN, a new wave of language change was on the horizon. Two programming languages appeared that were destined to have a profound effect on computer programming. These languages were C and Pascal. However, I did not seriously pursue either of these languages initially and so I'll defer my comments relating to them until a bit later. Instead, I became interested in a language that embodied a distinctly different paradigm — SNOBOL — and this is the direction I pursue next.

4.4 Advancing the Theory of Programming Languages

As briefly alluded to earlier, in the early 1960s a variety of formalisms for language description began developing that would have profound effects on the design and implementation of programming languages and systems. The regular expression concept from automata theory had already undergone substantial theoretical investigation, and had obvious potential for practical application. The use of BNF to describe ALGOL 60 instantly established it as a mainstay for programming language syntax. The Chomsky hierarchy of grammars and languages devised for the analysis of natural languages had immediate relevance to programming languages. And lastly, the development of the programming language

SNOBOL provided a cogent practical alternative for describing collections of strings.

Regular expressions first made their way into text editors and operating systems [318] and evolved into a substantially more expressive mechanism than the theoretical expressions that Kleene [203] had devised. This mechanism together with common extensions has no universally accepted name, but is often referred to as extended regular expressions and that is the term I will use in this chapter. Consequently, the well developed theory of regular expressions from automata research has limited applicability to the extended regular expressions that have come to be part of numerous programming systems. These enhanced mechanisms are a key component in a long list of modern programming languages, editors and systems [144].

Chomsky's work on grammars [68] provided a rigorous framework for formal language research and this blossomed into an active research area. I followed the theoretical side of the research on formal grammars diligently as it progressed. One of the grammars in the Chomsky hierarchy, the context-free (or type 2) grammar turned out to be a direct match with BNF, and its formal basis applied immediately to languages described by BNF. As a result, a variety of theoretical properties, plus ideas of direct practical use such as syntax trees and parsing methods were soon being adapted to programming languages (e.g., [141]).

At the heart of the SNOBOL language [123] was a construct known as a "pattern" that manifested a distinct alternative to describing and processing strings. The SNOBOL pattern concept anticipated the enhancements to regular expressions that would make them so practically useful, and included several more features of significant practical value. The aim of SNOBOL was practical rather than theoretical and practical programming ideas preceded any theory. But it was a relationship with other theoretical investigations I've noted that drew my interest, and I'll discuss these connections next.

SNOBOL was begun in 1962 by a group at Bell Telephone Laboratories led by David Farber. It went through a sequence of versions [165] before a widely available implementation of SNOBOL4 [166] became available for the IBM 360 System in 1968 when it attracted my serious attention. SNOBOL4 was intended to facilitate string processing, and provided a drastically different programming paradigm. The central feature of the language was the string pattern and the closely associated pattern matching process. By providing access, processing, and reuse of substrings encountered during the matching process, SNOBOL enabled succinct and flexible string processing in a way not found in any other language. Moreover, SNOBOL provided a clean and intuitive syntax, and embedded it in a success-failure control structure with automatic backtracking that was very natural to this processing paradigm. The emphasis on string processing did not lead to the exclusion of basic features of a general purpose language. SNOBOL4 included integer and floating point arithmetic, arrays, tables, and programmer defined data types. By 1970 I was using SNOBOL4 in an upper-level undergraduate course on concepts in programming languages.

The semantics of SNOBOL4 patterns were intimately connected to the matching process, and the success or failure of matching guided the flow of control with automatic backtracking in the matching process. The matching process was described in an entirely operational way [167] by an elaborate description of the scanning process and how each pattern element moved a cursor through a subject string. This included pattern elements that stored a partially matched substring in a program variable for later use (in the same pattern). Nonetheless, this paradigm had the potential for a much more declarative view. A basic aspect of a SNOBOL4 pattern was that it constituted the description of a collection of strings. It seemed of eminent value to me to gain insight into the limits of the kind of collection that might be describable through various types of pattern elements, and this motivated me to develop a formalism suitable for theoretical study of the pattern concept.

In 1971 I published a paper [130] that began with a formalization of the linked-list concept familiar from IPL V. Using this formalism I showed that one natural view of the sequencing of the elements appearing in a list structure was captured by the language of a context-free grammar whose structure is effectively isomorphic to that of the list structure. This formalism was then extended to capture SNOBOL4 patterns that include embedded defined functions and their application to matched (sub)strings. With this extended model, and using only finite-state string mappings for embedded functions, another result showed that any recursively enumerable language could be described. I found such theoretical insights to be of considerable practical value in programming problems involving pattern design, both through better general understanding and in the identification of pattern elements to be employed. This line of investigation of declarative semantics of patterns became a long-range research topic for me and resulted in several more papers in subsequent years [132, 139, 229]. In passing, I note that not until recent years has analogous theoretical investigation of extended regular expressions been seriously pursued (e.g., [9, 58, 143]).

Niklaus Wirth had joined the IFIP Working Group to define a successor to ALGOL in 1964. He eventually submitted one of two proposals to the group, but the competing proposal was selected. Wirth's proposal had included Hoare's suggestion [177] for dynamic records as an ALGOL extension. After this rejection by the IFIP Working Group, Wirth and Hoare independently pursued this proposal and produced a language that became known as ALGOL W [345]. This language was only implemented on IBM 360/370 computers, but I did briefly use it in our advanced undergraduate concepts course. ALGOL W was interesting for its development of types. The original FORTRAN [188] did not have types — it distinguished only fixed point and floating point values and variables. ALGOL 60 did incorporate the type concept, but was nearly as limited with only 'integer', 'real', and 'Boolean' declaration of variables. The ALGOL W proposal expanded the basic types of ALGOL 60 with 'complex', 'bit', 'string', and 'reference' types, and added a dynamic structured record type intended to accommodate list-structure programming provided earlier by languages such as IPL V. ALGOL W therefore provided a significant advance in giving type concepts a central role in programming languages, providing a basis for what we next see in Pascal.

By the latter half of the 1960s a movement to "structured programming" was beginning to gain broad support (e.g., [40, 109, 110]). A primary ingredient of this methodology was the use of single-entry/single-exit control features. Pascal was the creation of Niklaus Wirth [343] and was first announced in 1970. The Pascal language provided strong support for structured programming with a simplified ALGOL 60 for-statement, and added while-statements, repeat-statements, and case statements. Also Pascal expanded the steps taken by ALGOL W on the role of types. The reference type (changed to 'pointer') was given broad applicability, ordinal and 'char' types were included as simple types, enumerated and subrange types were added as was a new set type. And Pascal added a means for programmer defined types. Pascal required type declarations for all program entities, and continued the strong typing philosophy in compilation to prevent an operation or procedure being applied to unintended data.

The diverse combination of control and data structures made Pascal an attractive language for a wide variety of applications. None-the-less it was rather slow to attract adoption in the states. In the U.S., one factor was overcoming a large base of established software available for commercial languages. Another factor was that the initial compiler created by Wirth was for CDC computers that were not that common in U.S. industry. The most significant factor in the success of Pascal was the availability of a free and easily portable implementation known as the P-code compiler [262]. Wirth had devised a virtual computer

```
var A: array[1..2] of integer;
var B: array[1..2] of integer;
var C: record x,y: integer end;
var D: record x,y: integer end;
```

Figure 4.4: Pascal declarations

and an associated compiler so that all that was required for another computer implementation was writing a P-code interpreter. While this led to slow execution times that impaired industrial adoptions, it did not significantly deter an academic enterprise, and a no cost implementation plus the language's attractive features were irresistible in that arena.

As its popularity spread and implementations proliferated, the call for a standard for Pascal gained much support. One central feature of ALGOL 60 and its descendants is the idea of block structure and its use as the natural scoping unit. A colleague and I had noticed as the variety of kinds of elements in Pascal had grown and interactions abounded, not all the language rules honored this aspect of block structure. We believed it would be easy to implement a resolution of the cases we had observed, and made a contribution to the standards discussion [24, 25]. The totality of types had become quite diverse with enumeration and subrange types, arrays that may be 'packed' or not, variant and normal records, a new set type, plus an added means for the programmer to define types. Throughout its literature (e.g., the Pascal "bible" [195]), the phrase "same type" is used with no elaboration when expressing strong typing requirements, and while the standard [189] gave greater care in describing "type compatibility," the "same type" phrase was carried forward in numerous places. The purpose of strong static typing serves efficiency of implementation. But just as well it avoids errors of applying inappropriate operations to data values as early as possible, and this is an essential characteristic of Pascal typing. I remained with concern that "correct" typing so at the heart of the language should be left with any possible room for interpretation. For instance, with the declarations shown in Figure 4.4, could it be that variables A an B have the "same type" while variables C and D do not?

Later, I pursued an alternative to type equivalence in Pascal. In mathematics when two things are equivalent (i.e., the same) there are basic expectations for this relationship and it seems those properties should apply for Pascal types as well. Thinking about type equivalence led me to an idea for a somewhat more flexible approach that would still be straightforward to implement. My approach correlated with a natural "subtype" idea, and I developed precise definitions and an associated type checking algorithm in [134].

Two important design goals for Pascal [344] had been to serve as a sound language for teaching, including teaching system programming, and to achieve a high level of efficiency for both the compiler and compiled programs. As the success and longevity of the language clearly establishes, these goals were not only attained but exceeded. In 1982 Pascal became the language used in the beginning two course sequence in programming at the University of Iowa, and it remained in that role for more than a decade! And Pascal was the programming language of choice at the beginning of Knuth's campaign for literate programming [207]. In 1984 Niklaus Wirth received the ACM Turing Award for his contributions to programming language development.

4.5 Distinctive Language Alternatives

Languages that adopt an unorthodox perspective on computing have often held an attraction for me. Of course, just getting off the beaten path isn't necessarily an accomplishment. But internalizing a truly different approach does open avenues to new thinking. When facing an unfamiliar programming task, having diverse options enhances finding good choices. SNOBOL is one such language that I have already discussed. In this section I discuss several others that have expanded my computing perspective.

Kenneth Iverson developed a "programming notation" while teaching in a graduate automatic data processing program at Harvard in the latter half of the 1950s. His notation was heavily based on mathematics and matrices, intended as "a tool for communication and exposition" of algorithms [122], and motivated by dissatisfaction with then current programming notations. In its early years, APL was used as a means to communicate between individuals rather than as a computer programming language and this was immediately apparent in its appearance. Although I did not pursue APL until much later, the first widely available documentation was published as the book *A Programming Language* in 1962 [192]. As a brief illustrative example that distinguishes it from a programming language, one statement appearing in the APL book (p. 44) is shown in Figure 4.5.

The APL language has at its heart the array data structure with no restrictions on dimensionality. It has operations to conveniently construct arrays, to adjust the number and range of dimensions, and

$$u_2 \Phi_{u_1}^{\overline{u_2}} \leftarrow u_2 \Phi_{u_0}^{u_2}$$

Figure 4.5: an APL assignment statement

many array-oriented operations. And the basic scalar operations have their definitions extended to apply equally well to arrays of any dimension. For instance, for two linear arrays $x = (x_1, x_2, \ldots, x_k)$ and $y = (y_1, y_2, \ldots, y_k)$, the expression $x + y$ yields the array of element-wise sums. In a conventional language one would need to introduce an index variable and write a loop with a suitable termination test to describe the same computation. APL programs through this and numerous other conventions are both remarkably more succinct yet still improve clarity. Although Iverson describes the name APL as an acronym for A Programming Language, when I think about its profusion of array facilities, it's Array Programming Language.

Iverson joined IBM in 1960 where he initially collaborated on using his language as a means to describe computer hardware. It was not until the mid-1960s that serious effort began on developing a computer implementation of APL. This was challenging for several reasons. The first problem was that APL used a variety of math symbols, Greek symbols, special marks, and 2-dimensional spacing. Resolving this involved the development of a new terminal based on the IBM Selectric typewriter, plus a special APL type ball that provided a tailored 88 character alphabet with many of the unique APL characters (but only upper-case letters). Even so the implementation would still require an overstrike procedure for some operations. And with all that, the "linearization" of the notation required language adjustments. But despite these changes, the implementation preserved the original sense of the language quite well. In 1968 IBM released a version of APL for its System 360 series of computers [121], and it was with this interactive system that I became familiar with APL.

The often mentioned "one-liner" phenomenon of APL is a topic that gets us directly to the quintessence of the language. This is an informally defined term that refers to the composition of a selection from the profusion of APL operators to form an expression that accomplishes a non-trivial computation. I'll illustrate what's meant with a single but informative example — computing the prime numbers less than a given bound. This is a common exercise in many languages. It is one I first did on the IBM 650 (but not this way), and I was intrigued to see there is a published history of this programming problem [53]. Imagine this program for a moment as written in your favorite language. The point is

$$(\sim T \in T \circ . \times T)\,/T \leftarrow 1 \downarrow \iota N$$

Figure 4.6: APL primes program

that this is one line of APL code shown in Figure 4.6. I present it here for an insight into APL, and as another instance of the distinct appearance of APL programs.

This is not the place for a detailed account of APL operators and syntax, but a short informal description of the primes program in Figure 4.6 does reveal the alternative view that APL fosters. Given integer N, this program (read right-to-left) first creates a vector of values 2 through N, from that it creates the 2-dimensional matrix of all products from this vector, then it next selects occurrences in the original vector of integers that appear in this matrix (these must therefore be those values which are products), and finally inverts that selection to obtain the list of primes! So, no (explicit) loops, tests, etc.; no quotients and tests of remainders, just the creation of arrays, multiplications and selections (each by a single operation), and the correctness is self-evident! This program conveys the sense of APL's earlier use for communication between individuals. Of course, this program clearly overlooks consideration of both processor and storage efficiency, a common criticism of APL.

In 1979, Iverson received the ACM Turing award for his genuinely innovative ideas in programming. But he had a noteworthy encore that deserves greater recognition than it has received. This requires a brief chronological fast forward, but it best fits here. In 1989, Iverson collaborated with Roger Hui in the development of a language for an ASCII-based descendant of APL known as J [185]. The special characters for operators were replaced by a systematic and well-conceived pairing of ASCII characters, and intrinsic APL facilities were carried forward into J. But in addition to selecting a syntactic translation for APL, the language evolved in significant ways. The APL idea of array shape was generalized in J, and the idea of array rank was extended to functions, providing numerous opportunities for both generalizing and simplifying array operations. Although APL was not a functional language, it had a substantial functional core and in J this functional subset is expanded and emphasized. And free implementations of J are available for all the common platforms [198]. I have been fascinated to find an implementation available for my smart phone, where J's continued propensity for "one-liners" is a perfect match for the small screen.

In the mid-1970s interest developed in "abstract data types" (ADTs). Actually this term was used to describe two related but quite distinct

```
Domains:   Stack and Item (not otherwise elaborated)
Operations:
  new:   Stack
              -- result is a new (empty) Stack
  push:   Stack × Item → Stack
              -- result has Item added to the top
  pop:   Stack → Stack
              -- result has the top item removed
  top:   Stack → Item
              -- result is the top Item of a Stack
Axioms (for all s ∈ Stack and i ∈ Item):
  pop(push(s,i)) = s
  top(push(s,i)) = i
```

Figure 4.7: Pushdown stack algebraic ADT

ideas. The first was concerned with programming language facilities to allow users to code "first class" data types that have equal footing with those native to the language (e.g., [196, 227, 347]). This was quickly followed by the development of a methodology for specifying the behavior of a data type (e.g., [154, 168, 228]). Specification efforts were concerned with program proving [178], and with descriptions of new data types suitable for that purpose, plus completely avoiding implementation assumptions. The algebraic abstract data type became the central approach and this area expanded rapidly in the latter part of the 1970s. I found it remarkable that without choosing a representation for data nor providing implementations for the operations, a clear description of essential properties was still attainable. The algebraic ADT required only names for relevant type domains, operations (i.e., functions with no side-effects) involved with their type characteristic, and equational axioms for operation outcomes. To augment this vague algebraic ADT outline, a token version of the standard example of the pushdown stack appears in Figure 4.7. Note that it provides a term to describe each Stack, and the last-in-first-out behavior of operations on a Stack can be deduced from the equations (for brevity we ignore the issue of $top(new)$).

As a specification device, algebraic ADTs proved to be remarkably effective for a vast variety of descriptions. This is another instance of a device that while not a programming language, has such a close connection that it is appropriate to include in the discussion. The adoption of algebraic ADTs for specifications soon led to another problem. For more complex examples, the creation of the specification is itself a challenging technical task. While a suitable specification provides a beneficial means to gauge the correctness of code, an erroneous specification can have the doubly harmful effect of leading to incorrect code without realizing it. Just as we need to debug code by running it, it was soon found that computer aids were needed to discover unintended faults in specifications. Hence several projects to develop systems to automate the verification of an algebraic ADT specification arose (e.g., [153, 169, 263]). These systems provided execution behavior given only the algebraic ADT and further blurred the line with programming languages. While such tailored systems have some clear advantages, I observed that a widely available programming system, SNOBOL4, already provided support suitable for this task. I wrote a paper [133] that developed the direct use of ADT equations as SNOBOL4 code that allowed for testing of a specification, and included suggestions for dealing with some subtle issues that can arise in certain examples.

Through their clustering of a set of operations in direct proximity to their data, abstract data type ideas (both facets) were a factor in the development of object-oriented (O-O) programming. But the first language with objected-oriented features, SIMULA [83], had anticipated this view by the mid-1960s. This language was devised to enhance ALGOL 60 with facilities to aid in the creation of simulation programs. It was quickly recognized that those ideas were of much more general use, and the next step was SIMULA 67 [82] which provided objects, classes/subclasses, and inheritance similar to those found in many O-O languages we see today. And the relation of SIMULA 67 to specification and proving ideas was soon developed by Hoare [179] with a continuation of his earlier program proving work. Although I had followed ALGOL 60 closely, I never did pursue programming in SIMULA. In 2001 Ole-Johan Dahl and Kristen Nygaard received the ACM Turing Award for their original development of object-oriented programming.

My first encounter with O-O programming was Smalltalk-80 [155]. Work on Smalltalk began in the early 1970s and it went through several versions. In his discussion of the history of Smalltalk, Alan Kay [201] relates a wide range of factors influencing its development. Kay mentions the Sketchpad system [313], ALGOL and EULER, and work on porting an implementation of SIMULA. But the other factors he includes range to the broader topics of a changing computing environment with personal handheld computers, high-resolution displays, and windows based systems. There are certainly a wide array of technical and social factors that have made O-O programming so prominent, and fostered the explosive development of new O-O languages that has persisted ever since.

In the Smalltalk-80 description [155] the authors write that "Smalltalk is based on a small number of concepts," but I have to disagree. They base their contention on a list of five key words in the Smalltalk vocabulary: object, message, class, instance, and method. In fact, each of those words itself embraces a number of concepts. The word "object" involves concepts of data storage, operations upon that data and associated clustering, and some objects involve such ideas as the file system. The word "message" involves concepts of activation, of expression, and communication through concepts of argument and transmission of arguments, etc. The authors themselves essentially concede that the claim is an exaggeration when they write: "These five words are defined in terms of each other, so it is almost as though the reader must know everything before knowing anything." But as the strategy for organizing the language, the unwavering adherence to the object-oriented methodology reflected in those five words does indeed lead to a remarkably homogeneous and coherent result. Another consequence is a simple, easy to learn syntax. The last component is a class library that provides well-conceived organization. Where program libraries in conventional languages are adjuncts of occasional use, the class library is integral to Smalltalk. It also encapsulates the primary source of complexity in the system and so allows much of that to be internalized gradually. In 2003 Alan Kay received the ACM Turing Award for his work on the development of Smalltalk.

My experience with Smalltalk-80 was mostly limited to using it as a teaching system. I began using it in an advanced undergraduate course on programming language concepts in the latter-1980s. The initial Smalltalk-80 implementations were only available for interactive platforms with high-resolution graphics, and these machines were not feasible in our classroom setting at the time. The system I initially used was Little Smalltalk developed by Timothy Budd [52] who in facing a similar circumstance, developed a (freely distributed) implementation for UNIX-based systems.[83] This system made concessions to run on text terminals, but the essential sense of the language was still evident. Much later I published a paper [137] that presented a means to enhance algebraic ADTs to provide specifications for object classes. In their usual role, algebraic ADTs model collections of functions — there's no shared state or side-effects. My idea retained the equational character, and based the specification on sequences of messages rather than individual messages. This captures the shared store at the same level of abstraction as functions, and provides modeling for a changing store with no prescribed configuration of instance variables, just behavior of methods. The paper illustrates that a specification given in this way can be used for the verification of varied implementations and storage configurations, using `Smalltalk` as the programming language.

4.6   Declarative Languages

The "declarative" adjective could be applied to numerous languages these days. I won't pursue a careful definition, but for me the key issue is the focus on results, and the key feature is variables as they're known in mathematics not conventional languages. The algebraic `ADT` is an example I've already mentioned, and in this section I discuss three programming languages I've had occasion to consider previously in a case study comparison [136].

Functional programs in the form of lambda expressions [69] were already there at the dawn of the theoretical foundation of computer science, preceding computer programming. Later `LISP` [238] was an early version of (impure) functional programming with a version of lambda expressions. While conceptually interesting to me, I found `LISP` syntactically unattractive and never became motivated to pursue it seriously. As functional programming evolved, it remained a programming niche for me that I never quite made time for. A strong interest in functional programming did not develop for me until reading John Backus's Turing award paper in 1978 [16]. I found his rationale and motivation for functional programming very persuasive. And the algebraic flavor of his `FP` language resonated with my earlier mathematics background. Both the conciseness and the reusability characteristics of functions in this setting were highly attractive. And it was striking that his advocacy for functional programming was combined with a strong condemnation of traditional languages for their defects, especially since the award was given for his fundamental work on these earlier languages.

For some time I followed this area. I developed a local implementation of FP suitable for experimentation, and employed it for several years in an advanced undergraduate class on programming concepts. My interest continued to develop as I explored the contrasts in this paradigm. I found the computations that could be expressed using FP's 'insert' operations to accomplish looping (with no explicit iteration or recursion) quite remarkable. Backus and his colleagues had noted this in their writing, but it seemed of even greater significance to me. This was the same operation known as reduction in APL, and the FP insert-based programs were reminiscent of APL one-liners. These FP programs employ conditionals and inserts, but avoid explicit loops and recursion. This

Figure 4.8: Participants in the UT Year of Programming, August 1987

gives them a "tree structure" with linear execution along each path, and they embody a certain essence of FP programming. Eventually I published a paper [135] that provided a theoretical basis that identifies the computational expressiveness of these "insert-based" programs, showing that every primitive recursive function could be programmed in this way. This is an extensive class of functions that in its day was regarded in theoretical literature as including much of practical computing. My interest in functional programming was significantly heightened the following year (1987) when I attended the Institute of Declarative Programming at the University of Texas (Austin), one of the University of Texas Year of Programming series. I was motivated to hear John Backus speaking about recent work by his group in functional programming [18]. But what actually most caught my attention in his talk were his highly complementary comments about David Turner's work in functional programming.

David Turner had developed a series of three functional languages, plus devising a novel implementation technique for functional programming [323]. I subsequently focused on Turner's most recent language Miranda.[84] This was a pure functional language but with extensive supporting features. I was impressed by the conceptual foundation for Miranda where a program consists of defined functions and data, applying functions to arguments, and that's it. Both arguments and results of functions can themselves be functions. And the syntax was elegant, providing pattern matching for clean integration of multi-case definitions, "off-sides" rules for flexible but clear multi-line definitions, etc. Miranda also includes the 'insert' operation of FP (or reduction

```
primesTo n = [p | p <- [2..n]; ~members prds p]
            where prds = [q*r | q,r <- [2..n]]
```

Figure 4.9: Miranda primes program

of APL), called 'fold' (two versions actually), providing the means of functional programming that I found so fascinating in FP. Miranda included a strong yet flexible static type system, an attribute I believe contributes significantly to avoiding errors in programs, and one that FP side stepped. However, Miranda maintains an ease of use by not requiring type declarations but leaving the implementation to perform type inference (another application of the unification algorithm) to determine types. Also included were facilities for both abstract and algebraic data type definitions. I found it intriguing to see ALGOL 60's idea of "call-by-name" parameter transmission carried to logical conclusion in every feature of Miranda via "lazy evaluation." I was further impressed by a simple and practical comment convention that provided an effective means to capture the spirit of literate programming right in the language.

In conclusion for Miranda, I'll (re)consider the prime number program that was presented earlier as an APL example (Figure 4.6). If we follow the method used in APL, the Miranda version is the function definition shown in Figure 4.9.

Miranda is not as succinct as APL as it lacks such a breadth of pre-defined operators. But this program still replaces explicit loops with a pre-defined list comprehension operation and list membership predicate, and retains the appearance (and clarity) of straight-line code (a "two-liner"). The correctness is as self-evident as for the APL code (more so I'd say), but it's the same method so the efficiency issues remain. The clarity and coherence of Miranda is remarkable. I regard it as the best language design I have encountered and still enjoy programming with it. I note in passing that a Miranda program was used in the case study paper mentioned at the beginning of this section, and in 1989 the execution of that program took 50 minutes on our departmental VAX 11/780 and now takes less than one second on my personal iMac.

This brings me to the last language I will discuss. Prolog was initially devised in 1972-73 in a cooperative effort led by Alain Colmerauer and Philippe Roussel [73] with a primary motivation of natural language processing. A more mature version with broader goals began wider distribution in the mid-1970s. However, I remained only vaguely aware of the area for the next decade. In 1965 Alan Robinson had published the celebrated unification algorithm and resolution principle [291]. This was a major advance in automatic theorem proving, and provided a basis for the creation of the Prolog programming language and the initiation of logic programming generally. Alan Robinson became the founding editor of the Journal of Logic Programming in 1984. In 1985 he spent a week visiting in our department and lecturing on logic programming (esp. Prolog). Alan was very generous with his time throughout his visit, and an entirely gracious guest. His presentations were inspiring, and his visit was directly responsible for the development of a long-term interest in logic programming by myself and several colleagues and graduate students.

Prolog's basis in logic and subsequent selection of computational elements as relations rather than functions was a generalization with a persuasive potential to enhance expressiveness. One program can provide the computations of several programs in conventional (or functional) languages. Prolog's reliance on deduction holds the promise of "smarter" programs. And I found Prolog's backtrack search strategy reminiscent of the SNOBOL pattern match procedure, and it therefore felt like meeting an old friend. So I was quickly won over to this language. Subsequently several faculty and graduate students began a regular seminar on logic programming where we read a wide range of the available literature. Before long logic programming became one of the topics in our advanced undergraduate course on programming language concepts. I recently found it interesting to note a comment by Alan Kay in his history of Smalltalk [201]: "It is a pity that we did not know about Prolog then or vice-versa; the combinations of the two languages done subsequently are quite intriguing."

Somewhat later a colleague and I proposed using logic programming as a means to enhance the attribute grammar mechanism for semantic description of programming languages [299]. Rather than taking attributes as data values associated with derivation tree nodes and describing them by functional semantic rules, we advocated using predicates as the attributes. Then the attributes become active rather than passive elements, and the semantic rules are clauses describing the properties of these predicates. This enhances expressiveness with the generality of relational programming, and logic variables provide bidirectional communication that allows tree traversal (synthesized vs. inherited attributes) to be effectively transparent. Moreover, Prolog support for DCG grammars allows the syntax rules to be coherently embedded.

In the latter part of my teaching career, I undertook the effort to use Prolog (with a strong emphasis on underlying logic) as the language in an "experimental" first programming course for incoming freshmen at the University of Iowa. While I was uncertain at the outset, this course proved to be a popular success with the students who took it. I taught the course for six years, and soon became convinced that this is a superior way for students to develop conceptual foundations prevalent in programming, and of value in a variety of languages that might be later pursued, while becoming familiar with a tool of lasting practical utility. I described the approach I took in [138]. I'll end with one last note in passing. A Prolog program was also used in that case study paper mentioned at the beginning of this section, and in 1989 the execution of that program took 7 minutes on our departmental VAX 11/780 and now takes less than one second on my personal iMac, and a preferred version of that program which would only abort for lack of sufficient space on the VAX 11/780 now completes with no difficulty on my personal iMac.

4.7 Summary and Conclusions

As my professional career progressed, my realization that programming languages are much more than just tools to express our preformed ideas continued to grow. As we internalize a programming language, its form and structure molds our thinking at that level of abstraction. Creating a program is a process that requires bridging from a higher-level problem domain to a precise description about how a computation is to unfold within the conception of our adopted language. That intellectual process cannot avoid being shaped by the ideas embodied in the language used for its expression. Of course, we may conceive of a computational approach to solving a problem at an abstraction level apart from any programming language. But when it comes to getting a computer to carry out our approach, we will think of its solution all over again with our internalized knowledge of the organization and facilities provided by a chosen programming language providing the basis at every step. The thoughts underlying both construction and verification of our program can only occur to us by virtue of the structure, organization, and properties of the language that we have assimilated. So it is not just the expression of ideas, but also the very formation of our computational ideas for which we can thank our programming language.

The evolution of programming languages has led to an increasing distance between the programmer and the details of a particular computer. While a universal Turing machine may be sufficient for any computation, it is through more artful languages that we achieve "better" programs. As has been long realized, the metrics for "better" programs may be highly varied and so the proliferation of programming languages will surely continue. This is a phenomenon to be celebrated, as I foresee no limit to the progress that may be achieved. However, it may be challenging to determine which of numerous new ideas are actually "better." This is where I believe that solid knowledge of what has occurred in the past can improve our judgment.

In conclusion, I summarize those languages whose learning had the most significant impact on my personal understanding and appreciation of programming languages. This leads me to highlight nine of the languages I have mentioned. Each of these languages illustrates that the use of the collection of elements it incorporates leads to a distinctive conception of the programming process. The experience with each of them had a predominant influence on my understanding, and I briefly mention my reasons for each selection. I believe that each of these languages warrants broad general recognition for its contributions to fundamental programming language ideas, and that such general recognition is of value to further progress. However, I express my selection of these nine languages based on the importance of their influence in my personal development:

- FORTRAN: as the most successful early example to raise the level of abstraction far above that of the particular computer to carry out a computation, and encompassing a majority of features found in higher-level languages for years to come

- ALGOL 60 and BNF: for the convincing presentation of a means to precisely describe a systematically coherent, general, and flexible syntax, for nested scoping constructs, and for embracing recursion

- EULER: for demonstrating a methodology for a description of semantics of traditional languages that is both formal and practical, and for demonstrating the practical utility of highly restricted (i.e., precedence) grammars

- APL and its ASCII-based descendant J: for extensive development of an array-oriented language that provides a conceptually distinct paradigm for computing

- SNOBOL4: for an alternative paradigm based on backtracking and success/failure control structures, text processing, and generalized regular expressions

- Smalltalk-80: as an exemplary object-oriented language

- FP: simplicity and power of functional programming, the vast extent of programs that can be written without either iteration or recursion, and the suitability of algebraic methods of reasoning about programs

- Miranda: elegant functional programming, strong static polymorphic typing with type inference, and lazy evaluation used to great advantage in enhancing expressiveness

- Prolog: versatility and practical power of logic programming and the relational paradigm, formal (i.e., DCG) grammars as a programming feature, great utility for prototyping, potential benefits as a first programming language.

5. A Gentle Introduction to a TITANIC TURING TALE

It did not come as a surprise to me when I received an email on October 3, 2016, explaining that my POPL paper 'Category Mistakes in Computer Science at Large' was officially rejected. Nor do I in any way wish to ridicule the peer review process, which I consider to be professional and honest. I firmly believe the reviewer, also known as the POPL Gatekeeper in this book, did a very good job in:

1. Justifiably preventing my unorthodox analysis from entering the POPL'17 proceedings. (POPL is an abbreviation for Principles of Programming Languages.)

2. Protecting me from making a fool of myself in an auditorium full of programming language experts.

It is not entirely clear, however, whether the POPL Gatekeeper fully grasped my analysis (which, again, does not mean that he or she should have accepted my paper for publication). It should also be remarked that there was a second reviewer who also rejected my work. The comments of the second referee were extremely brief and I shall therefore not discuss them in the present book.

I start with the following timeline of events:

1. In the fall of 2015 I submitted a prior version of my work on category mistakes to the Communications of the ACM (CACM).

2. The CACM paper was rejected on January 24, 2016.

3. I used several CACM review comments in the next version of my work, which I submitted to POPL in the first week of July 2016.

4. On September 14, 2016, I received two peer reviews, each with the recommendation to reject my paper.

The reviews are very good and informative.

I am willing to believe the first reviewer when he says that my findings are already well understood in the POPL community. It should be remarked though that the CACM community, by contrast, fundamentally disagrees with my "category mistakes" and thus also with the reviewer's views. What to do about this?

Moreover, multiple POPL members still tell me that it *is* possible to fully verify a digital *system*, even after having studied my work in detail. So the reviewer's claim about the POPL community does not mix well with my daily experience. Please tell me what I can do about this? Should I abandon this research topic?

I do take gentle issue with the way the first reviewer frames my works: I am not presenting "anecdotes" nor am I just citing a few papers. Clearly my reference to Parnas and Chaudhuri backs up *my* story, not the reviewer's.

Figure 5.1: My rebuttal, written on September 14, 2016.

I'm not sure why CACM reviewers would ignore the difference between real-world systems and their mathematical models. I don't actually see that mistake in the quoted reviews. The standard convention is that a reference to a programming language refers to the abstract mathematical object, which is unproblematic, since today it is routine to define full-fledged languages unambiguously.

I don't think the Parnas-Chaudhuri exchange is problematic, because Parnas is not a member of the POPL community and likely ran into similar problems interpreting papers from it.

Figure 5.2: Reply by the POPL Gatekeeper on October 3, 2016.

5. On that same day I wrote and submitted a short rebuttal.

6. On October 3, 2016, I received the official rejection of my POPL paper, along with a reply from the POPL Gatekeeper pertaining to my rebuttal.

My rebuttal is shown in Figure 5.1 and the Gatekeeper's reply follows in Figure 5.2. There are four brief points that I wish to make with regard to both figures. First, the Gatekeeper is quite right in stressing that "today it is routine to define full-fledged languages unambiguously" and perhaps

I mis-portrayed (albeit in just one or two sentences) the accomplishments of the POPL community in my rejected paper.

Second, I am afraid that reviewers of the CACM have applied faulty reasoning and I have illustrated precisely this, both in my rejected POPL paper and again in Section 1.3 of the present book. A full account will follow in the next chapter.

Third, the reviewer sidesteps my observation in Figure 5.1 that "multiple POPL members still tell me that it *is* possible to fully verify a digital *system*, even after having studied my work in detail." I therefore suspect that the reviewer has not understood my paper very well after all. Of course, that is mostly if not entirely my mistake. I am, after all, still struggling to find the right words and tone to get my message across to the POPL community.

Fourth, reference to a programming language as an abstract mathematical object is "unproblematic" indeed; however, so is my insistence on consistently distinguishing between:

- computer programs and computer programming languages on the one hand, and

- mathematical programs and mathematical programming languages on the other hand.

If it later turns out that this distinction is superfluous, then nothing is lost and conceptual simplicity will have been gained. That is, I may want to conclude with the recommendation to fuse computer programs and mathematical programs (and, likewise, computer programming languages and mathematical programming languages), although I obviously do not wish to start my analysis with that recommendation.

It will turn out that the aforementioned distinctions *are* necessary for the sake of conceptual consistency. It will therefore also help to state these distinctions explicitly and frequently in this book. At this point I can already illustrate the distinctions in the context of C computer programming. A C program residing electronically in my laptop is a *computer* program. It can be mathematically modeled in several ways, and each of these ways results in a *mathematical* program. Likewise, the C *computer* programming language, as described by the C standard, can be formalized in multiple ways, resulting in various *mathematical* programming languages.[85] Assuming by convention that there is a one-to-one interdependence between the C computer program and some well-chosen mathematical program is fine; but using that premise to project theoretical results — and particularly impossibility results from computability theory — onto engineering is often misleading and sometimes plain wrong. I've illustrated this in Section 1.2 with Michael Hicks's exposition. More examples follow in the present chapter and in the next.

Finally, the Gatekeeper's implicit statement in Figure 5.2 that Parnas and I are non-members of the POPL community will be scrutinized in Section 5.2. First, I turn to another big name in formal verification: Andrew Appel.

5.1   Andrew Appel and Deep Specification

Conflations abound, and not only between categories but also between different objects belonging to the same category. To illustrate the latter, I will now comment on Andrew Appel's "the science of deep specification." It is only after having read Appel et al.'s on-line research statement [13] about this topic a few months ago that I became 100% convinced that most researchers in program verification have not taken the simple objections put forth by Brian Cantwell Smith [303], James Fetzer [125], Timothy Colburn [72], Donald MacKenzie [233], and other 'POPL outsiders' seriously.

To be precise, Appel et al. often refer to a computer program when they are actually referring to a *program text*, i.e., another representation of a̲ mathematical model of *the* computer program at hand. Figure 5.3 visualizes the current discussion.[86]

To explain Figure 5.3, first recall that I use the term "mathematical program" as a synonym for "mathematical model of a computer program." Second, note that a computer program can represent a mathematical program, as can the text of a program albeit in a very different manner. The former resides electronically in the computer at hand; the latter is printed on paper (or is projected onto your screen). For example, consider the following program text:

```
(defun factorial (n)
    (if (= n 0) 1
        (* n (factorial (- n 1)))
    )
)
```

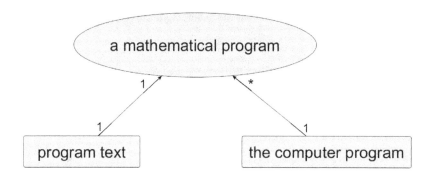

Figure 5.3: The computer program under scrutiny can be modeled using one or more mathematical programs. One such model is depicted, along with its program-text representation on, say, paper. (For simplicity, I assume in the present book that each mathematical program has precisely one program text and vice versa, but one is free to discard this assumption and thereby open the can of worms — i.e., the can of category mistakes — even further. A similar remark holds for the following over-simplification: each mathematical program has precisely one computer-program representation.) Specifically, the left arrow denotes the relationship 'is the textual representation of' and the right arrow denotes the relationship 'is the electronic representation of.'

This program text represents a mathematical object, i.e., a mathematical program. (One can perhaps also say that the program text represents the computer program. I certainly take no issue with this view, although in order to avoid confusion, I will not use the verb "to represent" in this manner.) One might be tempted to say that the program text *is* the computer program. This is frequently not a problem, although it is in the current foundational setting where one is encouraged, if not expected, to be as precise as possible.

The program text and the computer program are two different representations of the mathematical model. The first representation is intended for humans to read and the second is intended for a computer to interpret. A computer program resides electronically in a specific computer and is what most of us:

- would like to get "correct" and

- have difficulty comprehending since it is an artefact engineered for a digital machine.

Turning now to Appel et al.'s research overview, these authors state that until recently,

> program verification logics have not been 'live:' they have been about models of the program. [13]

However, all *mathematical* endeavors, including program verification, can at best be about mathematical *models*. Of course one can mathematically model a running computer program by resorting to, say, an operational semantics (of the mathematical programming language at hand). But how can mathematics become 'live'? What, exactly, is this supposed to mean?

Continuing with Appel et al.'s words:

> Program verification logics are now "connected directly and mechanically to the program". [13]

Two remarks can be made about this statement. On the one hand, Appel's first use of the word "program" in the previous quote must refer to a mathematical program, for how can one prove a logical statement without having a semantic model of the engineered computer program? Or, to be even more precise, the first occurrence of "program" refers to a textual representation of Appel et al.'s chosen mathematical program; that is, a program text. On the other hand, the second occurrence of the same word "program" in the previous quote refers to an electronic — humanly unreadable — representation, i.e., a computer program.

The take-away message, then, is that Appel et al. conflate their textual representation of their mathematical program (i.e., the program text) and the electronic representation of that same mathematical program (i.e., the computer program that we want to get "correct").

In more general terms, Appel et al. are essentially saying what Hicks has said before, namely that full verification of a hardware/software system is possible, both in principle and in practice. My claim is that this is not possible in principle but that program verification nevertheless does have a lot to offer in practice, especially if formal methodists would get their terminology right when disseminating their research findings to, say, software engineers in industry.

Continuing with the words of Appel et al.:

> we envision this network of verified software components, connected by deep specifications—specifications that are *rich, live, formal*, and *two-sided*. [13]

Some of these italicized words I do not understand at all, although perhaps this is my shortcoming. I am, after all, a 'POPL outsider.' Continuing:

> whatever properties are proved about a program will actually hold when the program runs [13]

This latter sentence exemplifies a category mistake *pur sang*, essentially already pointed out by James Fetzer in 1988 [125] and refined by me as follows: one can mathematically prove a *mathematical* property of a mathematical model of a computer program. One cannot prove a mathematical property of a computer program, regardless of whether the latter is running or not. Thus, the first occurrence of "program" in the previous quote refers to a mathematical program, while the latter occurrence — "the program runs" — must refer to a computer program.

To conclude, Appel and his colleagues fuse the category of mathematical objects (which includes mathematical programs) with the category of engineered artefacts (which includes computer programs). Perhaps there are very good reasons to do so and I am merely demonstrating my ignorance in this entire book. In any case, the 'science of deep specification' remains intriguing to me because I am not even able to grasp some of its basic principles.

5.2 The Parnas-Chaudhuri Exchange

The Gatekeeper has stated that Parnas and I are non-members of the POPL community (which is okay) and therefore we have misinterpreted — and continue to misinterpret — the work of Chaudhuri and other POPL members (which is not okay). From a sociological angle, the Gatekeeper's rather impressive remark fits right into the 2004 book *Mechanizing Proof: Computing, Risk, and Trust* [233], written by the sociologist Donald MacKenzie. Instead, the Gatekeeper could welcome in the name of the POPL community the scrutiny of Parnas and like-minded engineers in order to prevent two large communities, software engineering and computer science, from drifting further apart.

In the next section I present an extract from my rejected POPL paper in which we discuss an exchange between Parnas and Chaudhuri et al. The word "we" refers to my former student Rosa Sterkenburg and myself. Sterkenburg contributed to the excerpt but carries no responsibility for the other sections in the rejected POPL paper.

Extract from "The Chaudhuri-Parnas Dialogue" in my Rejected POPL Paper

To already [further] illustrate some of the issues just raised, we briefly turn to two recent writings of Swarat Chaudhuri [63, 64]. Chaudhuri has studied properties of programs that are necessary conditions for a correct program to possess in his mathematical framework. Starting with his 2012 article 'Continuity and Robustness of Programs' [64], Chaudhuri and his co-authors, Gulwani and Lublinerman, discuss a way to prove mathematical continuity of programs. A continuous program is not per se correct, but a correct program needs to be robust, and therefore, according to the authors it needs to be continuous.

[...] To put it candidly, Chaudhuri et al. do not seem to be aware that they are not proving something about their programs, but that they are proving something about some model of their programs. This is very clearly pointed out by David Parnas in a reaction to their article. Parnas says:

> Rather than verify properties of an actual program, it examined models of programs. Models often have properties real mechanisms do not have, and it is possible to verify the correctness of a model of a program even if the actual program will fail. [271]

Parnas specifically describes what he thinks is problematic about the mathematical assumptions underlying Chaudhuri et al.'s model:

> The article ignored the problem by both declaring: '...our reals are infinite-precision' and not specifying upper and lower bounds for integers. [271]

Because of these abstractions, [...] Parnas concludes in a similar way to [Brian Cantwell Smith] in 1985. Proving something [is] correct does not mean it will do what you want. "Some programs," Parnas continues,

> can be shown to be continuous by Chaudhuri's method but will exhibit discontinuous behavior when executed. [271]

Chaudhuri et al. reply to this critique as follows:

> From a purely mathematical perspective, any function between discrete spaces is continuous, so all *computer* programs are continuous. [271, our emphasis]

This reaction suggests that Chaudhuri et al. are not aware that considering a computer program as a mathematical function is a modeling step. A computer program is — and this will become our main conceptual point — categorically different from a mathematical program, even if the latter relies solely on finite abstractions. That is, a C program residing in a laptop (i.e., a computer program) is categorically distinct from its mathematical model, even if the semantics is "real semantics," i.e., a "detailed and cumbersome low-level model" of C. [...]

Pluralism Please

That, then, was a first extract from my rejected POPL paper. In retrospect, surely Chaudhuri et al. grasp the difference between a computer program and a mathematical program. (After all, who would not be able to make such a simple distinction?) But as you can judge for yourself it is Parnas who seeks conceptual clarity and who *does* understand what the POPL community is doing. Parnas is asking Chaudhuri et al. to be more precise in the way they formulate their research findings and not to oversell their mathematical results. It is Parnas who wants to differentiate between the "computer program" and its mathematical models — a.k.a. mathematical programs in the present book — while Chaudhuri et al. apparently do not want to comply with this.

The Parnas-Chaudhuri exchange is essentially a clash between definitions 1. and 3., presented in Section 1.4 and again in the following list, which contains four possible meanings for the term "computer program:"

1. a physical object à la Maurice Wilkes in 1950 and Dave Parnas in 2012,

2. a mathematical object of finite capacity à la Edsger Dijkstra in 1973,

3. a mathematical (Turing-machine) object of infinite size à la Christopher Strachey in 1973, and

4. a model of the real world that is not a logico-mathematical construction à la Peter Naur in 1985 and Michael Jackson today.

Grasping the multitude of definitions of a "computer program" results in an increased understanding of seemingly conflicting views on computer science: Parnas and Chaudhuri et al. use very different definitions for the term "computer program" and that is why the exchange of ideas between both camps is severely limited.

Appreciating and accepting the plurality of "computer program" definitions leads to better engineering practices. Consider for instance Byron Cook's article 'Proving Program Termination' [74] and Daniel Kroening and Ofer Strichman's book, *Decision Procedures: An Algorithmic Point of View* [214]. Together, these sources present two complementary views on what a "computer program" entails in the present century. For Cook, the variables in the following program text range over integers with infinite precision; this therefore follows Strachey's 1973 view of programming and shows some resemblance with Minsky's account on page 153 in his 1967 book *Computation: Finite and Infinite Machines* [245].

```
x : = input();
y : = input();
while x > 0 and y > 0 do
    if input() = 1 then x : = x - 1;
    else y : = y - 1; fi
done
```

Kroening and Strichman, by contrast, present an alternative perspective in which all program variables (in the above program text) are defined over finite-width integers; this aligns more closely with the view held by Dijkstra (1973) and with Minsky's account on page 25 in his 1967 book. The implication is that the software tools built by Cook differ in fundamental ways from those developed by Kroening and Strichman. (A decent analysis of Peter Naur's writings will reveal yet another view on computer programming.) Good engineers today benefit from this pluralism by using the tools provided from *both* aforementioned camps. They will not object if they are requested to use *different* semantic models for the *same* C computer program — a point that I have tried to stress before in Chapter 1.

5.3   Marvin Minsky and the Gatekeeper

Having presented several contemporary examples of category mistakes, I will now focus on a major historical actor: Marvin Minsky. The next subsection is another extract from my rejected POPL paper. It is about Minsky's reception of the halting problem in 1967. Again, it is the categorical distinction between a mathematical program and a computer program that I wish to emphasize, not to mention the fact that a computer program can be modeled with multiple mathematical programs. Minsky places a "computation system" and a "Turing machine" in the same category. That is fine as long as a computation system denotes a mathematical object. But, later on it becomes clear that he is talking about electronic computers as well.

Extract from "Marvin Minsky, 1967" in my Rejected POPL Paper

That all being said and done, the anonymous referees [from the CACM] whom I have cited at length [in a previous part of my rejected POPL paper] are in good company for they have history on their side. In 1967 Marvin Minsky reasoned similarly in his influential book *Computation: Finite and Infinite Machines*, as the following excerpt illustrates:

> The unsolvability of the halting problem can be similarly demonstrated for any computation system (rather than just Turing machines) which can suitably manipulate data and interpret them as instructions. In particular, it is impossible to devise a uniform procedure, or computer **program**, which can look at any computer **program** and decide whether or not that **program** will ever terminate. [245, p.153, my emphasis]

How do we interpret Minsky's wordings? Does a "program" solely refer to a "Turing machine" or any other Turing-equivalent mathematical program for that matter? If so, then we could perceive the statement as [...] abstract and correct; for then it would be a rephrasing of Martin Davis's 1958 account [86, p.70] of Alan Turing's 1936 paper. Or, as a second interpretation, does a "program" refer to a computer program that we can indeed "devise" and that may or may not "terminate"?

Based on a detailed study of Minsky's book, I assert that the right answer leans towards the second interpretation, as also the follow-up excerpt from his book indicates:

> (This observation holds only for **programs** in computers with essentially unlimited secondary storage, since otherwise the computer is a finite-state machine and then the halting

> problem is in fact solvable, at least in principle.) [245, p.153, my emphasis]

Paraphrasing Minsky, if we view computer programs as Turing machines, then we have an impossibility result with severe practical implications. However, Minsky's reasoning is only valid if the following two assumptions hold (each of which is wrong):

1. A computer program is a synonym for a mathematical program.

2. The mathematical program (mentioned in the previous sentence) must be equivalent to a Turing machine program and not to, say, a primitive recursive function.

In other words, Minsky did distinguish between finite and infinite objects, but not between abstract objects (Turing machines) and concrete physical objects (computers and storage), [...]

The second assumption often goes unmentioned in the literature exactly because computer programs and mathematical programs are frequently conflated. Contrary to what Minsky wrote, a computer with a finite memory is not a finite state machine, it can only be modeled as such.

Revisiting the Gatekeeper

That, then, was yet another excerpt from my rejected POPL paper and hence also from Chapter 6. The extract conveys Minsky's 1967 stance towards the unsolvability of the halting problem. The question of interest here is whether Minsky intentionally fused the two categories of mathematical programs and computer programs. Perhaps he was merely crafting his sentences with brevity in mind.

My scrutiny of the writings of Michael Hicks, Andrew Appel, Dave Parnas, Swarat Chaudhuri, CACM reviewers, and Marvin Minsky, strongly suggests that epistemic pluralism is part and parcel of computer science as we know it today, i.e., there is no one correct grasp, let alone a correct mathematical grasp, of software technology. It is precisely Raymond Turner — whose work I have yet to introduce — and like-minded scholars who are starting to lay the philosophical foundations of our emerging field. The POPL Gatekeeper, however, has expressed a very different sentiment in his review of my rejected paper:

> Most criticisms of particular authors and their published claims seem dubious. I think these authors really do

understand all the relevant issues and have just crafted certain sentences with brevity in mind. It doesn't work in practice to footnote every sentence with a review of all of its philosophical foundations!

It works to be precise at least once. Conflations abound. To borrow from Timothy Colburn's analysis, here is what Tony Hoare wrote in 1986:

> Computer programs are mathematical expressions. They describe, with unprecedented precision and in the most minute detail, the behavior, intended or unintended, of the computer on which they are executed.

This quote, which I have copied from page 132 in Colburn's book *Philosophy and Computer Science* [72], is reminiscent of Chaudhuri's words which Parnas and subsequently I, too, have scrutinized. Or, to use Peter Naur's 1989 words, as cited by Colburn on page 147 in his book:

> It is curious to observe how the authors in this field, who in the formal aspects of their work require painstaking demonstration and proof, in the informal aspects are satisfied with subjective claims that have not the slightest support, neither in argument nor in verifiable evidence. Surely common sense will indicate that such a manner is scientifically unacceptable.

One such subjective claim is once again that "full formal verification is possible," regardless of what philosophers and other scholars have had to say about this topic for several decades.

My point is that if the POPL Gatekeeper were to "really understand all the relevant issues" raised in my rejected POPL paper, then he would either agree with me *or* with Hicks, Appel, and others, but not with both camps. The Gatekeeper cannot have it both ways.

5.4 Albert Meyer, Dennis Ritchie, and the Gatekeeper

Category mistakes and incomputability claims seem to go hand in hand. I have given two examples in Chapter 1, one pertaining to Michael Hicks's views and another directly concerning the diagonal argument as perceived by an anonymous CACM referee. Now I shall give yet another example, also based on an excerpt from my rejected POPL paper. Specifically, I shall focus on the "code improvement problem," as discussed by Albert R. Meyer and Dennis M. Ritchie in their 1967 paper 'The complexity of loop programs' [243]. First, however, I present some terminology.

Terminology

I distinguish between two abstractions:

- abstraction A_I^{pos}, which allows for the representation of arbitrary large positive integers, and

- abstraction A_R, which allows for the representation of real numbers of infinite precision.

I use the term "programming language" as an abbreviation for "computer programming language" and always write "mathematical programming language" in full. The promised excerpt follows below.

Extract from "Flawed Incomputability Claims" in my Rejected POPL Paper

Let us turn to what I take to be flawed incomputability claims in the literature. Meyer & Ritchie's 1967 paper, for instance, is not solely about FORTRAN programs and LOOP programs, but, more generally, about the industrial "code improvement problem." Is it possible to automatically improve the assembly code of a corresponding FORTRAN program? — that's the central question in the introduction of their paper. Citing from their introduction:

> [C]onsider the problem of improving assembly code. Compilers for languages like FORTRAN and MAD typically check the code of an assembled **program** for obvious inefficiencies — say, two "clear and add" instructions in a row — and then produce edited **programs** which are shorter and faster than the original. [243, p.465, my emphasis]

As this excerpt shows, the authors specifically referred to the programming languages FORTRAN and MAD. So, clearly, a "program" here refers to a computer program, not to a mathematical object.

To get a mathematical grip on the industrial "code improvement problem," Meyer & Ritchie resorted to the theory of computability (and the theory of primitive recursive functions in the rest of their paper). Continuing with their introduction:

> From the theory of computability one can conclude quite properly that no code improving algorithm can work all the time. There is always a **program** which can be improved in ways that no particular improvement algorithm can detect, so no such algorithm can be perfect. [243, p.465, my emphasis]

Here we see Meyer & Ritchie refer to the undecidability of the halting problem. The implication is that a "program" now refers to a mathematical object and, specifically, to a "Turing machine" (see the second footnote in their paper for a confirmation). In other words, a "program" here does not refer to, say, a finite state machine, and it definitely does not refer to a program residing in a computer.

Meyer & Ritchie subsequently conveyed some common wisdom from the theory of computation. It is possible to decide the halting problem for specific subsets of Turing machines;[87] that is, even though the halting problem is undecidable in general, modeling computer programs as Turing machines can still lead to practical tools. (A modern example in this context is the work of Byron Cook et al. [74], which accepts both A_I^{pos} and A_R.) In their own words, regarding their code improvement problem:

> But the non-existence of a perfect algorithm is not much of an obstacle in the practical problem of finding an algorithm to improve large classes of common **programs**. [243, p.465, my emphasis]

Here the word "program" thus still refers to a "Turing machine."

My analysis so far shows that Meyer & Ritchie conflated their computer programs and their mathematical programs. Furthermore, Meyer & Ritchie only considered models of computation that comply with fundamental abstraction A_I^{pos}, namely that any variable V can represent an arbitrary large positive integer. Specifically, they only considered a mathematical program as synonymous [with] either a Turing-machine program (i.e., a partially computable function) or a LOOP program (i.e., a primitive recursive function). But, as we have seen, there are also models of computation that do not comply with abstraction A_I^{pos}.

Based on their mathematical analysis, Meyer & Ritchie ended their paper with an impossibility claim about computing practice:

> If an "improvement" of the code of a **program** is defined as a reduction in depth of loops or number of instructions (without an increase in running time), then the proof of [the undecidability result expressed in] Theorem 6 also reveals that there can be no perfect code improvement algorithm. Thus the code improvement problem, which we noted in the introduction was undecidable for **programs** in general, is still undecidable for the more restricted class of Loop **programs**. [243, p.469, my emphasis]

A purely mathematical — and, hence, valid — interpretation of Meyer & Ritchie's findings would amount to stating that the theoretical variant of the "code improvement problem" is not only undecidable for Turing-machine programs in general, but also for the more restricted class of LOOP programs. My disagreement has to do with the conflation made between the theoretical variant of the "code improvement problem" and the actual, industrial "code improvement problem." It is the latter which is "noted in the introduction" of their paper in connection with FORTRAN and MAD, *not* the former.

As a result then — and now a specific technical contribution of my analysis follows — it is still very well possible that somebody *does* end up defining the improvement of the code of a FORTRAN program as a reduction in depth of loops or number of instructions without an increase in running time and yet still obtains a perfect code improvement algorithm for these computer programs. (The adjective "perfect" can, however, only be quantified [with]in a well-defined mathematical setting.) Meyer & Ritchie's paper only indicates that the person who aspires [to] doing so will have to resort to a model of computation that does not comply with abstraction A_I^{pos}.

Michelle Obama Comes to the Rescue

That, then, was another excerpt from my rejected POPL paper. Perhaps the POPL Gatekeeper agrees with my analysis — because he says it "is only a recapitulation of dialogues familiar to everyone working in verification and related areas" — and perhaps he therefore considers my specific technical contribution to be trivial. I believe my analysis is very trivial indeed, but only in retrospect. Meyer and Ritchie support the following incorrect claim: one can mathematically prove that certain systems cannot be engineered. One main thread in this book is, after all, that constructing such a meta-claim amounts to making a category mistake (which requires rectification).

I honestly don't know whether the POPL Gatekeeper is *really* convinced that Meyer and Ritchie have incorrectly projected their theoretical findings onto programming practice. Here's what the Gatekeeper has to say in his own words about my rejected paper:

> I believe the main points of this paper are already widely agreed-upon in the POPL community and are used appropriately throughout the literature, even in the (relatively recent) papers that are quoted as evidence to the contrary. What I expect is that the author of this paper is an outsider to the POPL community and has misunderstood its conventions for writing about formalism.

The POPL Gatekeeper has made a similar remark about Parnas, who had the audacity to scrutinize the writings of Chaudhuri and some other honorable POPL members in the Communications of the ACM.

> [Continued:] The conclusions may be interesting as a data point on how outsiders may misinterpret formal-methods papers, though it's not clear the situation here is any worse than it is for the average technical field.

I have three things to say about this. First, is the Gatekeeper suggesting that I research whether "the situation" is worse in computer science (or formal methods in particular) than in other disciplines and that I then come back to the POPL gate if (and only if) the answer is affirmative? My educated guess is that researchers in more mature disciplines make fewer category mistakes. At least the Gatekeeper is implicitly acknowledging that "the situation" can be improved. That's a start.

Second, a researcher who does consistently distinguish between the model and the object that is being modeled is always preferable to someone who conflates the two; examine the Parnas-Chaudhuri exchange and judge for yourself.

Third, I'm afraid I'm not the kind of outsider the Gatekeeper was expecting. I talk to formal methodists every day. I even used to be a symbol chauvinist, but at some point in my career I met people who were asking the right questions.

Continuing with the Gatekeeper's words:

> I think this paper should not be accepted, because it is only a recapitulation of dialogues familiar to everyone working in verification and related areas, and yet at the same time the paper fails to "do no harm," taking quotes out of context to accuse researchers of confusion.

As Michelle Obama said recently, "when they go low, we go high."[88]

> [Continued:] As a convincing paper should, this one presents evidence for its thesis that the credo above is *not* already in wide use, explaining how research papers are giving misleading impressions about what their results prove. The evidence is entirely in the form of historical anecdotes and quotes from a handful of papers.

Historical anecdotes? A handful of papers? Is the Gatekeeper suggesting that I first find many more papers (which really presents no difficulty at all) and that I then come back to knock on the POPL gate? What about all the *other* examples presented in the books of Timothy Colburn and Donald MacKenzie, books cited in my paper? Moreover, almost all articles discussed in my rejected POPL paper are authored by prominent computer scientists.

> [Continued:] There are conventions for how such papers are written and what is assumed about their contexts. The author of this paper seems to misunderstand the conventions, leading to sightings of fundamental confusions where none exist.

Fortunately I belong to a second or even a third generation of software scholars who do not buy this rhetoric. I stand on the shoulders of James Fetzer, Peter Naur, Timothy Colburn and others who have already made complaints similar to those that I am making here. My small contribution, then, is that I am also

1. discussing technical claims pertaining to computability theory,

2. using primarily POPL case studies in an attempt to reach out to the POPL community, and

3. following Raymond Turner's very recent publications about technical artefacts (see later).

If only I could pass that POPL gate for a brief visit *without* having to accept the "conventions" of Hicks, Appel, Chaudhuri, et al.

Continuing with the Gatekeeper's words:

> At most, these new observations call for considering *edu-cational* strategies for informing a broader audience about these conventions; researchers in the field are already well aware of them and draw the correct conclusions from papers.

So now the Gatekeeper is at least admitting something. My clarifications are potentially beneficial to educators and outsiders; i.e., researchers who do not belong to the elite. (Thus, this book definitely has *an* audience.) I am afraid that the last part of the previous quote is false and, as I've explained before, I don't have any reason to believe that the POPL Gatekeeper actually understands the implications of consistently distinguishing between computer programs and mathematical programs either. Remember, he has sidestepped my observation that multiple POPL members still tell me that it *is* possible to fully verify a digital *system*.

5.5   Raymond Turner and Technical Artefacts

Until now I have merely distinguished between concrete physical objects and abstract objects. This trivial and generally accepted categorical distinction is all I need and all I have used so far to pinpoint flaws in everyday computer science research. If the reader is not convinced by now that rectifications are needed, then he or she will not become convinced while reading the rest of this book either.

All I have done so far is to place mathematical programs, finite state machines, Turing machines, and the like in the category of abstract objects; I have placed computers (including laptops and iPads) in the category of concrete physical objects. Furthermore, I have distinguished between a mathematical programming language, which is definitely an abstract object, and a computer programming language. The remaining question is: where do computer programs and computer programming languages belong? In the category of concrete physical objects? Or in a new, third, category of objects, called *technical artefacts*?

The short answer is to place computer programs and corresponding languages in the category of concrete physical objects. In fact, the reader can do precisely *that* throughout the entire book *without* undermining my main complaint. The more thoughtful response, however, is to acknowledge some recent advances in the philosophy of technology. Specifically, I will try to follow Raymond Turner's characterization of a computer program and a computer programming language as technical artefacts [325].[89]

Following one modern line in the philosophy of technology, and based on the writings of Maarten Franssen et al. [142], Turner brings a third category of technical artefacts to the table. Technical artefacts are intentionally produced things that can only fulfill their intended function because of their actual physical structure. Technical artefacts have a dual nature, determined by two sets of properties: functional properties and structural properties [325]. As Turner clarifies:

> Functional properties are articulated as black box specifica-
> tions in which the object of design is specified only in terms
> of its input and output behavior. [...] On the other hand,
> structural properties pertain to its physical makeup. [...]
>
> In summary, technical artefacts are individuated by the two
> descriptions: the functional and structural. The physical
> thing by itself is not a technical artefact. And of course there
> can be many such implementations. [...]
>
> [E]ach functional description and implementation determines
> a different artifact. [325, p.379]

The time has now come to turn to my entire rejected POPL paper in the next chapter.

6. TITANIC TURING TALE about Turing Completeness and Formal Verification

This chapter is a modified and improved version of an article written by the present author that has been peer reviewed and subsequently rejected by anonymous POPL referees in 2016.

Based on several case studies, I show that computer scientists have a tendency to slide in their discourse between *concrete physical* objects (laptops), *technical artefacts* (computer programs & computer programming languages), and *abstract* objects (Turing machines and the lambda calculus), as if there is no distinction between them. Specifically, several programmers (including myself) have asserted, but not proved, that computer programming languages X, Y, Z are Turing complete and have then used the alleged Turing completeness of such a technical artefact X to end up with an impossibility theorem about programming practice. I argue that this kind of reasoning is fundamentally flawed. Proving that some problem cannot be solved with a Turing-complete language (such as the undecidability of the halting problem of *mathematical* programs) does not imply that a related, industrial problem cannot be solved in practice (such as the termination problem of *computer* programs). Similar category mistakes will be illustrated throughout this chapter with regard to formal verification.

6.1   Introduction

In his 1985 paper 'The limits of correctness' [303], Brian Cantwell Smith discussed an important issue concerning program verification. He pointed out that a program always deals with the world through a model, it cannot deal with the world directly. There always is a layer of interpretation in place. Therefore, if we define a correct program as a program that does what we intended it to do, as opposed to what we formally specified it to do, proving a program correct is intrinsically very difficult.[90]

To illustrate the difficulty underlying Smith's *world-model* relation, assume we have a mathematical model of the world in which all apples are modeled as green and all strawberries as red. With this model it is quite simple to distinguish an apple from a strawberry, simply use the color to differentiate between the two kinds of fruit. If we thus write a program that can discriminate between red and green, then, at least with respect to our model, we can mathematically prove that our program can indeed distinguish between apples and strawberries. However, in the real world there exist red apples as well. So our model is not perfect; our proven correct program will tell us a red apple is a strawberry.[91] As Smith put it:

> Just because a program is 'proven correct', you cannot be sure that it will do what you intend. [303, p.18]

Smith argued that it is necessary for our model of the world, or *any* mathematical model for that matter, to be incorrect in some ways. For, if a model were to capture everything, it would be too complex to work with [303, p.20-21]. In the apple and strawberry example, the model need not capture the molecular structure of the fruits; doing so would presumably over-complicate matters for the application at hand. We need models to abstract away irrelevant — and, unfortunately, they also often abstract away relevant — aspects of the world in order to make an industrial problem mathematically tractable.

More important for the present chapter is the similar, yet slightly different, *programming-model* relation; that is, the relation between a computer programming language and the computer programs written in that language on the one hand and their mathematical counterparts on the other hand. This is a relation that Smith did not address in his paper. When proving a property about a computer program, such as deadlock prevention, some mathematical model of that program is used instead of the computer program itself. As a result, our formal proof of the computer program tells us something about it's model, and only indirectly something about the computer program itself. Here, too, we have the problem that our model is not perfect for it relies on abstractions from the engineered world.

On the one hand, it remains an open problem to find an ideal semantic model for, say, the C computer programming language. (Naively assuming it is even possible in principle.) On the other hand, and following the POPL Gatekeeper's useful feedback, today it is routine to define full-fledged *mathematical* programming languages, such as Standard ML. These two observations do not contradict each other. This chapter, and by extension this book, is more about C than about Standard ML.

Influenced by what is stated in the computer programming language's manual and how we choose to mathematically model the computer programming language at hand, two possible — and very different — key abstractions could be:

- abstraction A_I^{pos}, which allows for the representation of *arbitrary large positive integers*, and

- abstraction A_R, which allows for the representation of *infinite precision real numbers*.[92]

Moreover, the language manual, as David Harel puts it nicely in his book *The Science of Computing*, contains "a wealth of information concerning both syntax and semantics, but" — once again — "the semantical information is usually *not* sufficient for the user to figure out exactly what will happen in each and every syntactically legal program" [174, p.52-53]. Defining a formal syntax of a computer programming language is not considered a major problem anymore since the advent of the Backus-Naur Form (BNF) notation, but finding adequate formal semantics of an already existing *computer* programming language (such as C) remains a vexing research topic.

In the present exposition a "mathematical model" of a computer program (or computer programming language) refers to both the formal syntax and a formal semantics of the computer program (or computer programming language) at hand, although the formal syntax will often go unmentioned. It should be noted that in model theory, by contrast, numerals and other expressions (cf. syntax) are *modeled by* the natural numbers and other mathematical objects (cf. semantics).

To recapitulate, even a well-chosen mathematical model of a C computer program is not perfect for its semantics relies on mathematical abstractions, i.e., idealizations from the real, engineered world. As a result, and this is an important point for the rest of the chapter, *showing that something cannot be done according to a mathematical model of the engineered artefact under scrutiny does not necessarily imply that the engineered counterpart cannot be accomplished in practice.* Of course, one's formal semantics might become generally accepted as 'the standard' definition of the C computer programming language, but that sociological accomplishment will not invalidate the italicized statement made in the previous sentence with regard to C.

The Chaudhuri-Parnas Dialogue

To further illustrate some of the issues just raised, we briefly turn to two recent writings of Swarat Chaudhuri [63, 64]. Chaudhuri has studied properties of programs that are necessary conditions for a correct program to possess in his mathematical framework. Starting with his 2012 article 'Continuity and Robustness of Programs' [64], Chaudhuri and his co-authors, Sumit Gulwani and Roberto Lublinerman, discuss a way to prove mathematical continuity of programs. A continuous program is not per se correct, but a correct program needs to be robust, and therefore, according to the authors it needs to be continuous.

Chaudhuri, Gulwani, and Lublinerman do not address the world-model relation nor the programming-model relation in their 2012 article. While the first relation does not seem very relevant to us either with regard to their article, the second relation is, we believe, significant. To put it candidly, Chaudhuri et al. do not seem to be aware that they are not proving something about their programs, but that they are proving something about some model of their programs. This is very clearly pointed out by David Parnas in a reaction to their article. Parnas says:

> Rather than verify properties of an actual program, it examined models of programs. Models often have properties real mechanisms do not have, and it is possible to verify the correctness of a model of a program even if the actual program will fail. [271]

Parnas specifically describes what he thinks is problematic about the mathematical assumptions underlying Chaudhuri et al.'s model:

> The article ignored the problem by both declaring: '...our reals are infinite-precision' and not specifying upper and lower bounds for integers. [271]

Because of these abstractions, which include what we have called abstractions A_R and A_I^{pos}, Parnas concludes in a similar way to Smith in 1985. Proving something is correct does not mean it will do what you want. "Some programs," Parnas continues,

> can be shown to be continuous by Chaudhuri's method but will exhibit discontinuous behavior when executed. [271]

Chaudhuri et al. reply to this critique as follows:

> From a purely mathematical perspective, any function between discrete spaces is continuous, so all *computer* programs are continuous. [271, our emphasis]

This reaction suggests that Chaudhuri et al. are not aware that considering a computer program as a mathematical function is a modeling step. A computer program is — and this will become our main conceptual point — categorically different from a mathematical program, even if the latter relies solely on finite abstractions. That is, a C program residing in a laptop (i.e., a computer program) is categorically distinct from its mathematical model, even if the semantics is "real semantics," i.e., a "detailed and cumbersome low-level model" of C. (The words just cited come from Harvey Tuch et al. [319] and we shall see in Section 6.5 whether Tuch et al. agree with our categorical distinction or whether they reason like Chaudhuri et al.)

In a later paper entitled 'Consistency Analysis of Decision-Making Programs' [63], Chaudhuri, Farzan, and Kincaid do, however, address some of the issues raised by Smith in 1985. They write:

> Applications in many areas of computing make discrete decisions under uncertainty, for reasons such as limited numerical precision in calculations and errors in sensor-derived inputs. [63, p.555]

> Since the understanding of nontrivial worlds is often partial, we do not expect our axioms to be complete in a mathematical sense. [63, p.557]

These words are related to the model-world relation, which was not addressed in Chaudhuri's paper about continuity. With this remark Chaudhuri acknowledges that modeling of the real world is needed to get things done on a computer. Chaudhuri also indicates that this understanding usually is not perfect, which is close to Smith's point that one can never aspire having a perfect mathematical model. That said, Chaudhuri does not address the stringent programming-model relation and, hence, does not respond to Parnas's criticism.

Following Raymond Turner's recent writings, I will consistently and explicitly distinguish between *concrete physical objects* such as a laptop, *technical artefacts* such as the computer programming language C and computer programs written in C, and *mathematical models* such as a finite state machine and a universal Turing machine (not to mention the lambda calculus).

From Turner's perspective, both a laptop and a finite state machine are finite objects, but the former is physical while the latter is mathematical. Likewise, equating a laptop with a universal Turing machine is problematic, not primarily because the former is finite and the latter is infinite, but because the former moves when you push it and smells when you burn it while the latter can neither be displaced nor destroyed.

6.2   Conceptual Clarity

So much for my short critique of Chaudhuri's work, alluding to the possibility that also programming language experts can benefit from the present analysis.

I shall now be more precise about my chosen terminology.   All unreferenced quotes presented below come from anonymous referees of the Communications of the ACM and convey a general tendency in computer science to slide in discourse between laptops, computer programs, and mathematical models (e.g. Turing machines).

Specifically, and in line with some recent developments in the philosophy of technology, a distinction can be made between three separate categories:

1. computers, including laptops, are concrete physical objects,

2. computer programs and also computer programming languages are *technical artefacts*, and

3. Turing machines, finite state machines, and prime numbers are abstract objects.[93]

Further clarifications are in order. A *computer program* refers to the class of programs in standard, often commercial, *computer programming languages* that can actually be compiled and run on a specific electronic device; i.e., a *computer*. A *technical artefact* is a physical structure with functional properties [215]. Computer programs and corresponding computer programming languages are technical artefacts in that they only fulfill their intended function because of their actual physical structure. The physical manifestations alone, however, are *not* technical artefacts [190, 325]. Computer scientists build technical artefacts (e.g., computer programs, data types, and the like) which enable them to "reflect and reason about them independently of any physical manifestation" [325]. Reflecting and reasoning are often done by resorting to mathematics. Specifically, a *mathematical program* is a mathematical model — containing a formal syntax and a formal semantics — of a computer program.

Oftentimes computer scientists refer to a computer program while they are actually referring to a mathematical program instead. To be more precise, they are actually referring to a program text of the mathematical program under scrutiny and not to the program residing electronically in their computer (called a computer program). When computer scientists refer to their mathematical model they often incorrectly think that they are also directly referring to the actual computer program, nor do they distinguish between different representations of the mathematical program, such as a program text and a computer program. Each representation is a technical artefact. A visualization of the present discussion appears in Figure 5.3 in the previous chapter.

Consider, for example, the following words of Klein et al.

> Although they seem to at least model the implementation level, they did not conduct a machine-checked proof directly on the C code. [204, Section 6]

Klein et al. seem to suggest — at least to 'POPL outsiders' — that working "directly on the C code" is synonymous for working directly with the C computer program that will be deployed in, say, a safety-critical system. To illustrate that this is not the case, let us first briefly analyze what a human-checked proof entails before scrutinizing the previous excerpt in full. When conducting a human-checked proof, a human works directly on a program text; that is, on a human-readable and solely textual representation of *a* mathematical model of *the* C computer program under scrutiny and *not* on the C computer program itself. The program text and the C computer program are two very different representations of the mathematical model. A machine-checked proof, in turn, merely mimics the aforementioned human behavior electronically and thus does not work directly on the C computer program itself either.[94]

For another example from the same authors, consider the following statement:

> This means that if a security property is proved in Hoare logic about the abstract model [...], refinement guarantees that the same property holds for the kernel source code. [204, Section 2.3]

A mathematical property can only hold on a mathematical object, not on a technical artefact such as source code. Klein et al. do not distinguish between the kernel source code inhabiting our world and the mathematical counterpart of the code, even though both are close in some intuitive sense.

As a last example from the same source, let us scrutinize the following claim made about modeling:

> [T]o model the semantics of our C subset precisely [...]. *Precisely* means that we treat C semantics, types, and memory model as the standard prescribes [...] [204, Section 4.3, original emphasis]

Even if a subset of C is modeled "precisely," the model still remains categorically different from the artefact that is being modeled.

The examples just presented merely serve to illustrate that categorical distinctions can be made. In the next sections I hope to convince the reader that these distinctions *should* be made in the interest of advancing our field. I shall now re-visit each raised point in greater detail, starting with the topic of computers.[95]

Computers

In 1936, Alan Turing defined a *mathematical model* of a human computing a number as a process that requires no intuition nor going beyond the tireless following of rules. A decade later, Turing and others began to use a recast version of that same abstract object — a universal Turing machine — as a model for a general-purpose computer. Today, however, computer scientists carelessly use a universal Turing machine as a *synonym* for a computer. For example, when I insist that a computer can at best be viewed as a practical *instantiation* (with significant limitations) of a Turing machine, I receive the following rebuttal:

> My laptop *is* a universal Turing machine, but its tape size is of course limited by the finiteness of human resources.

Placing finite bounds on an abstract object (Turing machine) does not make it a concrete physical object (laptop). Instead, it results in another abstract object (e.g., a linear bounded automaton or a finite state machine) that can potentially serve as another model for the physical object at hand.

It is the mathematical model of a laptop and the corresponding mathematical modeling language that may or may not be Turing universal and Turing complete, respectively. A laptop cannot be any of these things.[96] Yet, many computer scientists disagree with this statement and erroneously place all objects in the same category. This is where a category mistake occurs. One can discriminate between

- finite and infinite mathematical objects (such as the numbers 3.14 and π), and between

- finite physical objects (laptops) and infinite physical objects (an example of which might be our universe).

But comparing a laptop with a Turing machine is only warranted with the proviso that we all agree we are reasoning across *separate* categories.

Computer Programs & Computer Programming Languages

Equally troubling is that several researchers merely assert, rather than debate, that 'a computer program and a computer programming language *are just as* a universal Turing machine, a recursively enumerable language, or a context-free language, and so on, in that *they are* mathematical models.' I shall present two reasons why this view is mistaken.[97]

Later I will play the computer scientist's advocate and assume for argument's sake that computer programming languages *are* abstract after all; i.e., that every computer programming language *is* a mathematical programming language. Inconsistencies will surface by the end of that discussion.

A first reason why a computer programming language is *not* just as a universal Turing machine or a linear bounded automaton follows naturally from the following simple question: How can the same computer programming language be a universal Turing machine for one researcher and a linear bounded automaton for another researcher? The answer, of course, is that a distinction has to be made between the computer programming language that we have engineered on the one hand and its models on the other hand. Today, some computer scientists model the C computer programming language with infinite-precision numbers. Others prefer a large, yet finite, state-space model (in which everything is decidable).

A second reason is that a computer programming language cannot inhabit our physical world in some sense and be an abstract object *at the same time*. The remedy, again, is to conceptually separate the computer programming language inhabiting our world from its mathematical models, leading to the disjoint categories of technical artefacts (computer programming languages) and abstract objects (mathematical programming languages), respectively.

Contrast my conceptual analysis, presented so far, with the following typical critique that I have received:

> The author claims that JAVA, C, and other programming languages can be modeled by a Turing machine but not compared to it.
>
> A programming language can indeed be compared to a Turing machine. When, for example, the programming language is defined to have bounds on its memory, it can be shown to be strictly weaker than a Turing machine, while in some other cases it can be shown to be Turing complete.

This line of thought is only correct if computer programming languages are abstract after all; i.e., if it is indeed the case that we need not distinguish between computer programming languages (such as JAVA and C) and mathematical programming languages. I will assume this to be so later. For now, I scrutinize the quote by noting, once again, that placing bounds on an abstract object (universal Turing machine) can only lead to another mathematical model (e.g., a linear bounded automaton) for the technical artefact (computer programming language) at hand.

The follow-up remark made by the same referee of the CACM goes as follows:

> There may be an issue with a programming language that is not completely defined. For example, leaving open the question of whether or not the memory is bound. In such cases, each specific implementation of the programming language, which *is still an abstract model*, might require a different handling.

While these words do not make sense philosophically for the same reasons presented above, they do however resonate when scrutinized by a historian. As Figure 6.1 illustrates, before the 1960s, the standard was machine-dependent "coding;" i.e., computer programs were viewed as concrete physical objects. During the 1960s, Corrado Böhm, Giuseppe Jacopini, Edsger Dijkstra, Christopher Strachey, and others gradually made a switch to machine-independent, ALGOL-like languages, leading to the academic birth of computer science [102, 317]. In this context people such as Dijkstra and Strachey viewed a computer programming language *as* a mathematical object, albeit in very different ways. Only recently have Nurbay Irmak, Raymond Turner, and others made progress in obtaining a more coherent terminology [190, 325] for computer science's most relevant terms: "computer program" and "computer programming language."

In the latter setting of Irmak and Turner, a "specific implementation" of a "programming language" is most definitely *not* an abstract model. Coming back, then, to the previous quote from the CACM referee, I thus take a "specific implementation" to refer to a computer programming language, *not* a mathematical programming language!

Implications for the Halting Problem

Viewing computers programs and computer programming languages as technical artefacts (or, more generally, as objects that are *not* abstract) has implications for computer science's most celebrated result: the unsolvability of the halting problem. Let us revisit Strachey's 1965 letter to the editor of the Computer Journal, entitled 'An Impossible Program' [309]. In his letter, Strachey presented a "proof" of what he called "a well-known piece of folklore among programmers," namely that

> [I]t is impossible to write a program which can examine any other program and tell, in every case, if it will terminate or get into a closed loop when it is run.

What is a Computer Program?

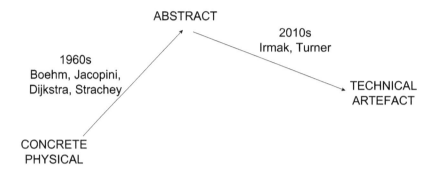

Figure 6.1: A computer program (and, likewise, a corresponding computer programming language) has been assigned to different categories during the course of history: from a *concrete physical* object to an *abstract* object and, recently, as a *technical artefact*. The latter interpretation has yet to be scrutinized let alone welcomed by programming language specialists, hence the purpose of writing the present chapter.

Strachey's programs were written in the Combined Programming Language (CPL), developed in Cambridge.

In his "proof," Strachey introduced $T[R]$ as a Boolean function taking a routine (or program) R with no formal or free variables as its argument such that, for any R,

- $T[R] = True$ if R terminates if run, and

- $T[R] = False$ if R does not terminate if run.

Strachey then examined the following routine:

```
rec routine P
    ♮L :      if T[P] go to L
              Return ♮
```

He noted that, if $T[P] = True$, then routine P will loop, and it will only terminate if $T[P] = False$. Therefore, in each case, $T[P]$ has exactly the wrong value and this contradicts the existence of the computable function T [98, p.29].

Strachey's line of reasoning appears in many textbooks. However, I stress again that considering a computer program as a partially computable function is a *modeling* step. Strachey could instead have followed some of his contemporaries by using primitive recursive functions [243] or finite state machines [112] to model his CPL programs (see Section 6.3) and then might have come to the conclusion that it *is* (trivially) possible to algorithmically analyze whether CPL programs halt.

Viewing computer programs and corresponding computer programming languages as technical artefacts (or, more generally, as objects that are *not* abstract) leads to the conclusion that Strachey's "proof" is *not* on par with proofs of statements that *are* purely mathematical, such as:

- There are infinitely many primes.

- The halting problem of Turing-machine *modeled* CPL programs is undecidable.

- The halting problem of finite state-machine *modeled* CPL programs is decidable.

Grasping the significance of the observations made so far is not easy for here is yet another response that I have received from a CACM reviewer:

> What does the undecidability proof of the halting problem for computer programs actually tell us. Like diagonalization proofs in general it may be viewed finitely as saying that, if there is a bound M on the size of accessible computer memory, or on the size of computer programs, or any other resource, then no computer program subject to the same resource bounds can solve the problem for all such computer programs.

The previous remark and the follow-up remark, presented below, are only correct if we accept the following two assumptions (both of which are wrong):

1. A computer program is a synonym for a mathematical program.

2. The mathematical program (mentioned in the previous sentence) must be equivalent to a Turing machine program (i.e., a partially computable function) and not to, say, a primitive recursive function.

The reason why the second assumption has to hold is merely because the computer scientist is referring to the halting problem of Turing machines. Continuing:

> If computer program A solves correctly all halting problems for computer programs respecting bound M, then the counterexample computer program T must exceed that bound, which is why A fails for T. To solve problems of computer programs, one needs an ideal program.

This quote hints at a distinction that must be made between finite and infinite objects (with the latter being labeled "ideal"); however, the categorical distinction between computer programs and mathematical programs goes completely unnoticed. Again, this is where a category mistake occurs. The undecidability *proof* of the halting problem concerns *mathematical* programs only, not *computer* programs.

Playing Computer Science's Advocate

I shall now assume that computer programming languages *are* abstract any way. Suppose thus that we can do away with the categorical distinction between a computer programming language and its mathematical counterpart (contrary to what some philosophers of technology and I advocate). We simply have one language L which is both mathematical and inhabiting our physical world at the same time. Then, since it is not uncommon for a computer programming language to have different meanings world-wide — e.g., in 1973, the same ALGOL-like computer programming languages were perceived mathematically as Turing complete by Strachey but not by Dijkstra (see Section 6.3) — it follows that our programming language L can have an undecidable halting problem for one research group and a decidable halting problem for another group. In principal, there is nothing wrong with this peculiar state of affairs. It has taken several centuries for the concept of a mathematical function to acquire one ubiquitous definition among mathematicians,[98] so the same might very well happen — or has already happened (think of Standard ML for example) — for the concept of a "programming language" as well.

However, even if it turns out that researchers across the world unanimously accept the definition of programming language L to be one well-specified Turing-complete language, it still remains erroneous to deduce the following popular statement made in computer science:

(*) No Turing machine can solve the halting problem of that program over there, residing in your laptop.

While language L might be abstract and any program P written in language L might be abstract too, each program P^* that resides in someone's laptop is most definitely not an abstract object: burning the laptop amounts to destroying that particular instance program P^*, while it is not possible to burn mathematical objects, such as P and L.

The correct statement should therefore go as follows:

A. No Turing machine can solve the halting problem of L programs.

And if the reader *does* accept the distinction between the three aforementioned categories (i.e., between concrete physical objects, technical artefacts, and abstract objects), along with my preference to place L programs in the category of technical artefacts, then the correct statement is:

B. No Turing machine can solve the halting problem of Turing-machine *modeled* L programs.

Observe that both Statements A and B are very similar. The reason is that, at some point, a distinction has to be made between *mathematical* programs and *computer* programs, irrespective of whether a programming language L is abstract or not. For consistency's sake, then, I shall continue to explicitly distinguish between computer programming languages and mathematical programming languages as well.[99]

The crux, thus, is that computer scientists have unjustified reasons to project an impossibility theorem (from the beautiful abstract world of computability theory) onto, say, C programs residing in laptops — as demonstrated by the previous quote (*). Doing so amounts to making a titanic category mistake.

Marvin Minsky, 1967

That all being said and done, the anonymous referees whom I have cited at length are in good company, for they have history on their side. In 1967 Marvin Minsky reasoned similarly in his influential book *Computation: Finite and Infinite Machines*, as the following excerpt illustrates:

> The unsolvability of the halting problem can be similarly demonstrated for any computation system (rather than just Turing machines) which can suitably manipulate data and interpret them as instructions.
>
> In particular, it is impossible to devise a uniform procedure, or computer **program**, which can look at any computer **program** and decide whether or not that **program** will ever terminate. [245, p.153, my emphasis]

How should we interpret Minsky's wordings? Does a "program" refer solely to a "Turing machine" or to any other Turing-equivalent mathematical program for that matter? If so, then we could perceive the statement as abstract and correct, for then it would be a rephrasing of Martin Davis's 1958 account [86, p.70] of Alan Turing's 1936 paper [320]. Alternatively, as a second interpretation, does a "program" refer to a computer program that we can indeed "devise" and that may or may not "terminate"? (As a third possibility, perhaps Minsky wanted to convey that the halting problem, which concerns abstract Turing-machine behavior only, is also unsolvable when engineers resort to their computer programs instead of mathematical programs.)

Based on a detailed study of Minsky's book, I assert that the right answer leans towards the second interpretation, as the follow-up excerpt from his book also indicates:

> (This observation holds only for **programs** in computers with essentially unlimited secondary storage, since otherwise the computer is a finite-state machine and then the halting problem is in fact solvable, at least in principle.) [245, p.153, my emphasis]

Paraphrasing Minsky, if we view computer programs as Turing machines, then we have an impossibility result with severe practical implications. However, Minsky's reasoning is only valid if the following two assumptions hold (both of which are wrong):

1. A computer program is a synonym for a mathematical program.

2. The mathematical program (mentioned in the previous sentence) must be equivalent to a Turing machine program and not to, say, a primitive recursive function.

In other words, Minsky did distinguish between finite and infinite objects, but not between abstract objects (Turing machines) and concrete physical objects (computers and storage), let alone between abstract objects and technical artefacts (computer programs).

The second assumption often goes unmentioned in the literature precisely because computer programs and mathematical programs are frequently conflated. Contrary to what Minsky wrote, a computer with a finite memory is *not* a finite state machine; it can only be *modeled* as such.

While I have criticized the first assumption on philosophical grounds, the second assumption can easily be scrutinized by taking a close look at the history of computer science. Specifically, I will complement the findings of the historians Michael Mahoney [234, p.133] and myself [210, Ch.3] and support the following thesis: computer scientists have mathematically modeled actual computations in diverse ways, and many of them have not resorted to the Turing-machine model of computation (or to any other Turing-equivalent model). In my words:

> There never has been a standard model of computation throughout the history of computer science; it is the quest for such a model and not the model itself that brought many computer scientists together.

After my historical survey (Section 6.3), I will zoom in on flawed claims about incomputability (Section 6.4) and formal verification (Section 6.5), followed by my conclusions (Section 6.6).

6.3   A Historical Mini-Tour

My tour will start with FORTRAN and ALGOL 60 and end with a few remarks about Dijkstra's work on predicate transformers and the early beginnings of model checking. To keep my account coherent, however, I will first discuss and then frequently refer to Martin Davis, Ron Sigal, and Elaine Weyuker's authoritative 1994 textbook, *Computability, Complexity, and Languages: Fundamentals of Theoretical Computer Science* [92].

Davis et al. note in their textbook that the concept of "algorithm" does not have a general definition that is separated from a particular language [92, p.69]. The word "algorithm" is informal in nature, while the word "language" refers here to an artificial and mathematically well-defined object. The authors provide the definition of such a little mathematical programming language $\mathcal{L}$ in the second chapter of their book, with each $\mathcal{L}$-program denoting a partially computable function.

More precisely, the authors define a partial function to be partially computable if it is computable by some $\mathcal{L}$-program [92, p.30]. Later on, the authors also introduce Church's Thesis to the uninitiated reader in the following manner: "any algorithm for computing on numbers can be carried out by a program of $\mathcal{L}$" [92, p.68-69].

For example, given the mathematical function $f(x_1, x_2) = x_1 + x_2$, the following $\mathcal{L}$-program [92, p.22] captures, in purely mathematical terms, how $x_1 + x_2$, denoted by y, could in principle be obtained from x_1 and x_2 in a practical, operational setting:

$$
\begin{array}{ll}
 & Y \leftarrow X_1 \\
 & Z \leftarrow X_2 \\
[B] & \text{IF } Z \neq 0 \text{ GOTO } A \\
 & \text{GOTO } E \\
[A] & Z \leftarrow Z - 1 \\
 & Y \leftarrow Y + 1 \\
 & \text{GOTO } B
\end{array}
$$

The convention to use capital letters (e.g., X_1, Y) in $\mathcal{L}$-programs and the precise mathematical meaning of each $\mathcal{L}$-instruction is irrelevant for understanding the present chapter, but they are of course spelled out in Davis et al.'s textbook.

More important are the following two observations. First, both the function f and the $\mathcal{L}$-program are mathematical models of a computer program P that is expressed in some common computer programming language.[100] Second, Davis et al. have chosen a Turing-complete language to model the common, yet unspecified, computer programming language.

We can thus, at least in principle, view the toy mathematical programming language $\mathcal{L}$ as one valid modeling technique for a computer programming language such as C. The modeling technique includes an operational semantics of language $\mathcal{L}$, since Davis et al. also provide a formal description of an abstract machine that interprets $\mathcal{L}$'s syntax. The abstract machine's behavior thus models the execution of program P residing in an actual computer.

1950s–1960s: FORTRAN

Davis et al.'s textbook is about computer science theory in general, and less about computer programming languages per se. Consider, therefore,

Albert Meyer and Dennis Ritchie's 1967 paper, 'The complexity of loop programs' [243], which is explicitly about FORTRAN computer programs on the one hand and their mathematical counterparts, called "LOOP programs," on the other hand. Here, however, the class of LOOP programs corresponds precisely with the class of primitive recursive functions, which, in turn, is strictly contained in the class of partially computable functions. The crux is that Meyer and Ritchie chose a mathematical programming language (i.e., their LOOP language) that is *not* Turing complete in order to model FORTRAN as it was known and used in the 1960s.

Meyer & Ritchie's choice for a Turing incomplete language to model FORTRAN computer programs seems reasonable if we keep in mind that the designers of FORTRAN had eschewed unrestricted (i.e., potentially unbounded) forms of recursion from their computer programming language in the 1950s and 1960s [98]. It would thus not have made much sense to model FORTRAN, as it was defined and used back then, with a Turing-complete mathematical language.

Comparing and contrasting the aforementioned writings of Davis, Sigal, and Weyuker on the one hand and Meyer & Ritchie on the other hand in greater detail would lead to the following two additional observations. First, both sources rely on a similar mathematical notion of *state* that captures the specific values of each variable V. Second, both modeling techniques rely on fundamental abstraction A_I^{pos} (namely that any variable V can represent an *arbitrary* large positive integer) but not on A_R. In Meyer & Ritchie's words:

> A Loop program is a finite sequence of instructions for changing non-negative integers stored in registers. There is no limit to the size of an integer which may be stored in a register, nor any limit to the number of registers to which a program may refer [. . .] [243, p.466]

Also noteworthy is that Meyer & Ritchie's "registers" denote mathematical objects, referring thus only indirectly to the real, physical registers of an actual computer.

Meyer & Ritchie's preference to model physical computations by resorting to the set of primitive recursive functions and not some other strict subset of all partially computable functions (cf. finite state machines), deserves more reflection though. Early versions of the FORTRAN computer programming language relied solely on the representation of variables that stand for numbers with a fixed, machine-dependent bound; it was impossible to represent arbitrary large integers and real numbers in FORTRAN computer programs. For example, in one version of FORTRAN, real numbers had to be represented in absolute value in between the lower bound 10^{-38} and the upper bound 10^{38} [279, p.50]. Meyer & Ritchie's 1967 abstractions underlying their mathematical LOOP programming language were thus partly at odds with the original machine-dependent FORTRAN programming system that had been developed in the 1950s.

It is not too difficult to find small conflations between the mathematical LOOP language and the engineered FORTRAN language. For example, when reasoning about the loop structure of a mathematical LOOP program, it is not only tempting to think about the actual execution of a corresponding FORTRAN computer program in a physical machine, but to also express the mathematical findings in terms of physical quantities. In Meyer & Ritchie's words:

> [C]omputation for inputs larger than two would not termi-
> nate during the lifetime of the Milky Way galaxy. [243]

The computation here refers not to a physical computation but to an abstract one; that is, to a particular behavior of LOOP's underlying abstract machine. Meyer & Ritchie thus expressed abstract computations not in terms of the number of, say, encountered mathematical instructions or mathematical states of computation, but in terms of the "lifetime of the Milky Way galaxy." Physical notions such as time and lifetime are, however, completely meaningless in the abstract.

Claiming that the previous wordings by Meyer & Ritchie merely resemble a *façon de parler* in the computer science arena will, as I shall show in Section 6.4, not do in general.

1960s: ALGOL

Stepping away from FORTRAN, let us now have a look at the ALGOL 60 computer programming language. It was officially defined in 1960 by a committee of American and European researchers, including several numerical analysts such as the Dane Peter Naur [98, Ch.7] and the Dutch man Adriaan van Wijngaarden [257, Sec.2.3]. In contrast to FORTRAN, ALGOL 60 *was* very recursive in nature and, in fact, re-known for that reason [8, 184]. Expressing potentially unbounded forms of recursion in an ALGOL 60 computer program was thus allowed. (The objective, then, was to have the runtime system raise an error if the unbounded nature of the computation would exceed the finite resources of the computing machine.) However, the fundamental abstraction A_R, which allows for variables V to denote infinitely precise real numbers, was, just like with FORTRAN, *not* admitted in ALGOL 60's official definition even though specific machine bounds such as 10^{-38} and 10^{38} were deliberately not mentioned in the official report [20].

To be more precise, let us examine Naur's intentions with ALGOL 60. "Numbers and variables of type real," he wrote in 1960, should be "interpreted in the sense of numerical analysis" and, specifically, "as entities defined inherently with only a finite accuracy." Naur elaborated:

> [T]he possibility of the occurrence of a finite deviation from the mathematically defined result in any arithmetic expression is explicitly understood. No exact arithmetic will be specified, however, and it is indeed understood that different hardware representations may evaluate arithmetic expressions differently. The control of the possible consequences of such differences must be carried out by the methods of numerical analysis. This control must be considered a part of the process to be described, and will therefore be expressed in terms of the language itself. [20, p.304-305]

In retrospect then, and not very surprisingly, ALGOL 60 was not as machine independent as various modern computer programming languages are today. ALGOL 60 was first and foremost intended to be a tool for numerical analysis; moreover, the shift from numerical computation to symbolic manipulation happened mostly during the 1960s and thus after ALGOL 60 had already been officially introduced [288].

Another researcher of standing was van Wijngaarden who contributed to the design of ALGOL 60 (and in even more significant ways to that of ALGOL 68). As we shall see, van Wijngaarden followed Naur's dictum of not abstracting away the finiteness of physical computing machinery when mathematically modeling computer arithmetic.

In his 1962 report 'Switching and Programming' [336, p.11], van Wijngaarden discussed the smaller-than relation ($<$). If two expressions $e1$ and $e2$ are equal ($e1 = e2$), he said, can then $e1 < e2$ be simplified to *false*? Van Wijngaarden responded to his own question as follows:

> This cannot be said without more. Actually in a well-known computer system hold both relations $-0.0 = 0.0$ and $-0.0 < 0.0$.

Observations of this kind led van Wijngaarden to start formalizing the properties that all his machines of interest had to share. In other words, he started developing a mathematical model of a generic computer. He defined his model by introducing axioms $A1$, $A2$, ..., $A11$ in his exposition [336, p.11].

> [W]e postulate the following axioms and definitions:
> $A1$: $e1 = e1$
> $A2$: $+e1 = e1$
> [...]
> $A11$: $e1 > e2 \land e2 > e3 \supset e1 > e3$

According to van Wijngaarden, his formalization discarded actual machines that did not satisfy one or more of his axioms, rather than — as I prefer to put it — discarding mathematical *models of* actual machines. For example, according to van Wijngaarden, the first axiom $A1$ does not hold for analogue computers. But what van Wijngaarden actually meant (or should have meant) is that $A1$ discards *models of* analogue computation.[101] (Note that most, if not all, computers *are* analogue in nature.)

Van Wijngaarden presented his axiomatic-based approach to a wide audience in Stockholm, August 1964, and this resulted in his 1966 paper 'Numerical Analysis as an Independent Science' [338] (which predates Tony Hoare's famous 1969 article, 'An Axiomatic Basis for Computer Programming' [178]). Numerical analysis, van Wijngaarden said in 1964, should be considered as an "independent science, i.e. independent of ordinary pure mathematics." Continuing:

> Of course, the concepts with which numerical analysis deals, can all be considered as approximations to concepts in pure mathematics. A numerical analyst may, in this sense, compute an approximation to a zero of a function, to a limit, to an integral, to a derivative, and so on. However, he might also consider the result of the computation to be the thing that he wanted to have, and the "mathematical" concepts as approximations to his "numerical" concepts. [338, p.66]

Van Wijngaarden emphasized that he did not want to make a science on "a particular make of apparatus," but neither did he want to view exact arithmetic as the ideal either; that is, he did not embrace abstractions A_I^{pos} and A_R.

To recapitulate, in the 1960s many computer scientists shared the research agenda of abstracting away from actual computers. But the ways in which to abstract — more precisely: the ways to *mathematically model* computational processes — were very diverse indeed.

It should also be noted that several computer scientists drastically changed their view on computer programming between the early 1950s and the 1970s. One prime example is Strachey who, like so many of his contemporaries, had heavily opposed recursion in the early 1960s but, by 1973, was promoting a very different approach [98, 311].

1970s: Strachey and Dijkstra

In 1973, Strachey wanted to view an ALGOL-like computer programming language, and ALGOL 60 in particular, *as* a mathematical object and of a particular kind. Strachey's line of reasoning went as follows: first ensure program correctness in an exact mathematical setting, before descending to the level of bit patterns inside the finite machine. Note, however, that this transition from an exact mathematical setting to a level of finite bits is not merely a transition from a world of infinite objects to a world of finite objects. It is also, and foremost, a transition from a category of abstract objects to *another* category of technical artefacts (or, if the reader prefers, a category of concrete physical objects).

The luxury of a high level language, Strachey continued, was that it allowed the programmer to reason about *variables* instead of *locations* (or *addresses*) and *functions* instead of *subroutines*. When an ALGOL 60 programmer writes a statement such as $x := Sin(y + 3)$ then he has in mind the mathematical *functions* sine and addition. For Strachey, the finiteness of the actual machine — which is categorically different from the finiteness of van Wijngaarden's abstract machine — only entered the picture afterward.

> It is true that our machines can only provide an approxima-
> tion to these functions but the discrepancies are generally
> small and we usually start by ignoring them. It is only after
> we have devised a program which would be correct if the
> functions used were the exact mathematical ones that we can
> start investigating the errors caused by the finite nature of
> our computer. [310, p.2]

In his paper, based on a close collaboration with the logician Dana Scott, Strachey assigned "the usual mathematical" properties to his domains: Booleans T, Integers N, Reals R, etc. Likewise, he defined stored values V as $T + N + R$ with the infinitely long number π belonging to R. He then formalized ALGOL 60 in terms of these domains — for example, he defined vectors as $L^* = L + L^2 + L^3 + \ldots$ where L denotes the domain of locations — and, hence, appropriated ALGOL 60 in this infinite setting [3, p.5,16,17]. Strachey subsequently used his modeling technique to compare and contrast a variety of computer programming languages, including ALGOL 60. Note, however, that this was all happening under the umbrella of exact mathematics. The finiteness was intentionally ignored and abstractions A_I^{pos} and A_R were embraced.

In contrast to Strachey's views, van Wijngaarden's former pupil Edsger Dijkstra viewed programming arithmetic and computation in general in a finite mathematical setting in 1973. Dijkstra, who had been educated in theoretical physics, expressed his dislike for the fundamental, logically flavored, abstractions that Strachey and Scott were advocating:

> We are considering finite computations only; therefore we can restrict ourselves to computational processes taking place in a finite state machine — although the possible number of states may be very, very large — and take the point of view that the net effect of the computation can be described by the transition from initial to final state. [112, p.1]

Dijkstra explicitly attributed the introduction of "infinite computations" to the logicians Turing and Scott [112, p.1] and continued as follows:

> Thanks to Cantor, Dedekind et al., however, we know that the inclusion of the infinite computation is not a logically painless affair, on the contrary! In the light of that experience it seems more effective to restrict oneself to finite computations taking place in a finite, but sufficiently large universe, thereby avoiding a number of otherwise self-inflicting pains.[112, p.1]

In retrospect, it would only be a slight exaggeration to say that Dijkstra mathematically modeled physical computations with finite state machines, even though his 1976 work on predicate transformers [113], for instance, was in accord with abstraction A_I^{pos} (but not with A_R). That is, arbitrary large positive integer values could be used as input to Dijkstra's ALGOL-like mathematical programs.

To better understand Dijkstra's views, two important observations are in order. First, contrary to Minsky and several other researchers who were reasoning in terms of à posteriori verification of mathematical programs, Dijkstra used the unsolvability of the halting problem to motivate that programming and correctness proving should go hand in hand. The programmer, Dijkstra said, should restrict his programs — and I say: his mathematical programs — so much that he can prove that they halt. His approach became one of the cornerstones of the field of activity called programming methodology [98, p.39].

Second, and similar to Davis et al.'s account, Dijkstra modeled the operation of an actual program instruction as a change of mathematical state. But, unlike Davis et al., each of Dijkstra's states resembles a set of assertions on data instead of the specific values of the data. Loosely speaking, Dijkstra's states are thus more general than those of Davis et al. and, moreover, finite in number. This explains Dijkstra's emphasis on finite state diagrams. Those diagrams, in turn, can be constructed on the basis of the program control structures — i.e., sequence, selection, and iteration — that he had already heavily advertised in his earlier work on structured programming [81, 111].

Dijkstra's predicate transformers influenced many computer scientists, including E. Allen Emerson who helped pioneer model checking (cf. [118, p.34]). The industrial success of model checking lies in the fact that several computer programs can be modeled "at a meaningful level of abstraction by a *finite* state system" [118, p.37, my emphasis]. Continuing with Emerson's recollections from 2008, "model checking," he says,

> validates in small part the seemingly naive expectation during the early days of computing that large problems could be solved by brute force computation including the technique of exhaustive search. [118, p.40]

Emerson's words and my historical observations lead to three take-away messages. First, the word "program" acquired very different mathematical meanings during the past decades. As a result, and this is my second point, it would be a drastic over-simplification to assert that a mathematical program can only be synonymous to a "Turing machine." Third, it would be equally naive to claim that a mathematical programming language has to be synonymous to a "Turing-complete language."

The three points just raised, along with my previous philosophical point not to conflate mathematical programs and computer programs, lead to the following important conclusion:

> Showing that some mathematical problem cannot be solved with a Turing-complete language (and thus also not with Turing-incomplete languages) does not imply that the related,

but dissimilar, industrial problem cannot be solved in practice.

Of course, I am primarily, but not only, alluding to the undecidability of the halting problem of *mathematical* programs on the one hand and the termination problem of *computer* programs on the other hand — as I shall now elaborate.

6.4   Flawed Incomputability Claims

Let us turn to what I take to be flawed incomputability claims in the literature. Meyer & Ritchie's 1967 paper, for instance, is not solely about FORTRAN computer programs and mathematical LOOP programs, but more generally about the industrial "code improvement problem." Is it possible to automatically improve the assembly code of a corresponding FORTRAN computer program?   — that is the central question in the introduction of their paper. Citing from their introduction:

> [C]onsider the problem of improving assembly code. Compilers for languages like FORTRAN and MAD typically check the code of an assembled **program** for obvious inefficiencies — say, two "clear and add" instructions in a row — and then produce edited **programs** which are shorter and faster than the original. [243, p.465, my emphasis]

As this excerpt shows, the authors specifically referred to the computer programming languages FORTRAN and MAD. Clearly, a "program" here refers to a computer program, not to a mathematical object.

To get a mathematical grip on the industrial "code improvement problem," Meyer & Ritchie resorted to the theory of computability (and the theory of primitive recursive functions in the rest of their paper). Continuing with their introduction:

> From the theory of computability one can conclude quite properly that no code improving algorithm can work all the time. There is always a **program** which can be improved in ways that no particular improvement algorithm can detect, so no such algorithm can be perfect. [243, p.465, my emphasis]

Here we see Meyer & Ritchie refer to the undecidability of the halting problem. The implication is that a "program" now refers to a mathematical object and, specifically, to a "Turing machine" (see the second footnote in their paper for a confirmation). In other words, a "program" here does not refer to, say, a finite state machine, and definitely does not refer to a program residing in a computer.

Meyer & Ritchie subsequently conveyed some common wisdom from the theory of computation. It is possible to decide the halting problem for specific subsets of Turing machines, that is, even though the halting problem is undecidable in general, modeling computer programs as Turing machines can still lead to practical tools. (A modern example in this context is the work of Byron Cook et al. [74], which accepts both A_I^{pos} and A_R.) In their own words, regarding their code improvement problem:

> But the non-existence of a perfect algorithm is not much of an obstacle in the practical problem of finding an algorithm to improve large classes of common **programs**. [243, p.465, my emphasis]

Here the word "program" thus still refers to a "Turing machine."

My analysis so far shows that Meyer & Ritchie conflated their computer programs and their mathematical programs. Furthermore, Meyer & Ritchie only considered models of computation that comply with fundamental abstraction A_I^{pos}, namely that any variable V can represent an arbitrary large positive integer. Specifically, they only considered a mathematical program as synonymous with either a Turing-machine program (i.e., a partially computable function) or a LOOP program (i.e., a primitive recursive function). But, as we have seen, there are also models of computation that do not comply with abstraction A_I^{pos}.

Based on their mathematical analysis, Meyer & Ritchie ended their paper with an impossibility claim about computing practice:

> If an "improvement" of the code of a **program** is defined as a reduction in depth of loops or number of instructions (without an increase in running time), then the proof of [the undecidability result expressed in] Theorem 6 also reveals that there can be no perfect code improvement algorithm. Thus the code improvement problem, which we noted in the introduction was undecidable for **programs** in general, is still undecidable for the more restricted class of Loop **programs**. [243, p.469, my emphasis]

A purely mathematical — and hence valid — interpretation of Meyer & Ritchie's findings would amount to stating that the theoretical variant of the "code improvement problem" is not only undecidable for Turing-machine programs in general, but also for the more restricted class of LOOP programs. My disagreement has to do with the conflation made between the theoretical variant of the "code improvement problem" and the actual, industrial "code improvement problem." It is the latter which is "noted in the introduction" of their paper in connection with FORTRAN and MAD, *not* the former.

As a result then — and now, a specific technical contribution of my analysis follows — it is still very possible that somebody *does* end up defining the improvement of the code of a FORTRAN computer program as a reduction in depth of loops or number of instructions without an increase in running time and yet still obtains a perfect code improvement algorithm for these computer programs. (The adjective "perfect" can, however, only be quantified within a well-defined mathematical setting.) Meyer & Ritchie's paper only indicates that the person who aspires to doing so will have to resort to a model of computation that does not comply with abstraction A_I^{pos}.

Jeroen Voeten's "Fundamental Limitations"

Jeroen Voeten's 2001 article 'On the fundamental limitations of trans-formational design' [333] is in many ways similar to Meyer & Ritchie's 1967 paper yet it is much more elaborate, both in theoretical content and in terms of impossibility claims. The rectification made in the previous paragraph holds here as well.

It is Voeten's article that triggered me a decade ago to start investigating the historical and philosophical underpinnings of computer science. I scrutinize Voeten's article in Appendix A and provide more research context in Chapter 8.

Fred Cohen's "Computer Viruses"

Now that we have trained our philosophical eye, it becomes rather easy to find similar flawed incomputability claims in the literature at large. My final example with regard to incomputability is primarily based on Fred Cohen's seminal 1987 paper, 'Computer Viruses: Theory and Experiments' [70].

Cohen's abstract reads as follows:

> This paper introduces "computer viruses" and examines
> their potential for causing widespread damage to computer
> systems. Basic theoretical results are presented, and the
> infeasibility of viral defense in large classes of systems is
> shown. [...] [70, p.22]

In his paper, Cohen mathematically *models* computer-virus programs as
Turing machines [70, p.25]. Subsequently, he uses the undecidability of
the halting problem to make a direct claim about industrial practice [70,
p.22], as the following quote from his paper indicates:

> Protection from denial of services requires the detection
> of halting programs which is well known to be undecid-
> able [146]. [70, p.22]

Strictly speaking, this line of reasoning is flawed. Protection from denial
of services "is undecidable" with Cohen's Turing-complete modeling
language, it is not impossible in an absolute, *practical* sense. Therefore, I
am tempted to re-phrase Cohen's abstract like this:

> In this paper we model "computer viruses" as Turing
> machines and subsequently use a classical undecidability
> result from computability theory to gain insights into the
> industrial problem of building viral defense systems. [70]

Later on in his paper, Cohen states that he has presented infeasibility
results that "are not operating system or implementation specific, but
are based on *the* fundamental properties of systems" [my emphasis].
Continuing, he writes that "they reflect realistic assumptions about
systems currently in use" [70, p.34]. However, it should be stressed,
once again, that the Turing-machine model of computation is *not the only
model* of computation that has been used throughout history in *attempts*
to achieve these research goals. Dijkstra, just to give one example, would
have preferred to use a large, yet finite, state space for his model of
computation.

Cohen frequently conflates industrial problems such as the "detection"
of computer viruses and theoretical notions such as "undecidability."
For example,

> Precise detection [of computer viruses] is undecidable,
> however, statistical methods may be used to limit undetected
> spreading either in time or in extent. [70, p.34]

It is in Cohen's model that precise detection of mathematically modeled
computer viruses is an undecidable problem, not outside the model.
Likewise,

> Several undecidable problems have been identified with
> respect to viruses and countermeasures. [70, p.34]

Cohen has identified several undecidable problems with regard to his
model of the industrial problem at hand. No more can be said.

A very similar and more recent account on computer viruses is Eric
Filiol's 2005 book, *Computer Viruses: from theory to applications* [126]. Filiol
begins the technical part of his exposition by putting "Turing machines"
front and center (which is fine), yet without acknowledging that he is, in
fact, selecting only one possible modeling technique (which is not fine).
In Filiol's words:

> The formalization of viral mechanisms makes heavy use of
> the concept of *Turing machines*. This is logical since computer
> viruses are nothing but computer programs with particular
> functionalities. [...]
>
> A Turing machine [...] is **the** abstract representation of what
> a computer is and of **the** programs that may be executed with
> it. [126, p.3, my emphasis]

I take issue with the boldfaced words. Filiol's reasoning is only valid
if we accept that computer programs can only be adequately modeled
with Turing machines. But accepting *that* premise would amount to
ignoring a large part of the history of computer science (not to mention
current-day practice). Alternative models of computation, such as linear
bounded automata, have been introduced in our field exactly because
the Turing-machine model was perceived as too baroque [234].

6.5   Formal Verification Unplugged

Stepping away from incomputability claims, let us now turn to a
2009 prize-winning paper of Klein et al. [204], which is about formal

verification of digital systems. Klein et al. provide a far more nuanced picture of the field than, say, Minsky in 1967. For instance, they write that:

> Verification can never be absolute; it always must make fundamental assumptions. [204, Section 2.3]

Several careful statements such as the previous one constitute the cited article. But let us now zoom in on the following excerpt:

> We assume the correctness of the compiler, assembly code, boot code, management of caches, and the hardware; we prove everything else. [204, Section 1]

Even if the authors had proved the hardware correct (by resorting to a mathematical model of the hardware), then this feat would not imply that the hardware behaves correctly [72, 233]. I presume Klein et al. are very well aware of this philosophical observation but perhaps not of all its profound implications. For example, consider the very first sentence of Klein et al.'s abstract:

> Complete formal verification is the only known way to guarantee that a *system* is free of programming errors. [204, my emphasis]

This statement is wrong and also seems to contradict the first quote, presented above. What formal verification can do in principle is guarantee that a mathematical model of the system is free of mathematically modeled programming errors. It cannot provide guarantees about the engineered system itself, even when the claim is restricted to programming errors only. Likewise, complete formal verification can definitely provide more confidence about system correctness, but it cannot provide guarantees about the system. (It should be mentioned that Klein et al. emphasize the word "guarantee" in connection with "formal" verification and that I am therefore reacting accordingly.) Continuing:

> [T]he kernel will never crash, and it will never perform an unsafe operation. [...] We can predict precisely how the kernel will behave in every possible situation." [204, Abstract]

It is in the model of the kernel that we will not encounter a crash in some appropriate mathematical sense, not outside the model. Moreover, it is according to the model that the kernel will never perform an unsafe operation. It is in the mathematical model that we can predict precisely how the kernel will behave in every possible situation that has been modeled.

Even if the model is a "low level" model of C, a categorical distinction has to be made between the mathematical model of the C computer program and the C computer program itself. The former does not inhabit the real world while the latter does. Guaranteeing that the kernel will never crash, as the authors emphasize in their paper [204, Section 6], is not what their research efforts have achieved. And, just to be clear, formal verification definitely has reasons to be pursued (in my opinion).

6.6   Closing Remarks

Several computer scientists truly believe that mathematics can capture reality perfectly. A real computer system, they say, *is* a mathematical model. That is, both a laptop and a program residing in the laptop can be treated mathematically without any loss, or without any significant loss, of information. According to Parnas, several computer scientists *do* realize that mathematics *cannot* capture reality perfectly *but* they just don't care about it. For them it is the mathematics that matters and the real world is "beneath" them.

It is of course true that mathematics allows us, amongst many other things, to model existing software/hardware systems so that we can better understand what we have built. But it is plain wrong to believe the following popular statement made in computer science:

> We can investigate a mathematical model of the system to *mathematically prove* that the system will always satisfy certain requirements.

We can prove that a model will satisfy certain requirements but not that the system will. If we find mistakes in our mathematical model, then this does *not necessarily* imply that the computer program under scrutiny (or, to give another example, a digital circuit) contains flaws. Likewise, if we show that something cannot be done according to a model, then this does *not exclude* the possibility that the engineered counterpart can be done in practice after all.

According to the philosophy of technology, claiming that some mathematical model *is* the real system is similar to claiming that the number 5 *is* my left hand. While my left hand can be modeled with the number 5 it cannot be replaced by it. Likewise, my laptop can be modeled with some mathematical machine model of computation, such as the universal Turing machine, but it cannot be replaced by it. A laptop is a *concrete physical* object while the mathematical machine model is an *abstract* object. A similar remark can and has been made in the present chapter about programs residing in laptops on the one hand and mathematical models such as the Turing machine on the other hand.

Computer science would be much better off if we were to consistently view mathematical results relative to the chosen model and, thus, *not to perceive our mathematical findings as absolute claims about the real world*. I therefore propose to improve the previous statement as follows:

> We can investigate a mathematical model of the system to *obtain extra confidence* that the system will always satisfy certain requirements.

Just like testing and debugging computer programs have limitations, verifying properties of a system's mathematical model has fundamental restrictions as well. So, on the one hand, the following statement made by Dijkstra is true:

> Program testing can be used very effectively to show the presence of bugs, but it is hopelessly inadequate for showing their absence.[102]

On the other hand, even mathematical techniques cannot guarantee the absence of real bugs. Proving properties on a mathematical model can at best show the absence of mathematically modeled bugs.[103]

Testing and debugging computer programs amounts to working with concrete physical objects (computers) and with what I have called *technical artefacts* (computer programs and program texts), while proving properties of a mathematical model amounts to reasoning about abstract objects. This difference is a bit superficial though because we also accomplish the latter by using program texts and other technical artefacts when conducting proofs.

In principle, then, formal verification can guarantee that a mathematical model of the system is indeed free of mathematically modeled bugs. In practice, however, one or more representations of the mathematical model are used to obtain that assurance and, furthermore, by carrying out a process that is intrinsically as error prone (in some informal sense) as testing and debugging.[104]

What I have shown in this chapter is that we, computer scientists, frequently make two mistakes. The first is philosophical and the second is historical in nature:

1. We have the tendency to place computer programs and mathematical programs in the same category.

2. Many of us think an adequate mathematical programming language has to be Turing complete.

Grasping these mistakes leads to the following technical insight: it makes no sense to argue for nor against the claim that "C is a Turing-complete language," as has been done in the literature at large.[105] Doing so amounts to making a category mistake. The computer programming language C can be modeled in different ways; i.e., with Turing-complete and with Turing-incomplete mathematical languages. Just like several different models can and should be used to gain multiple insights about my left hand, engineers and academics alike could perhaps appreciate more the multitude of semantic models that exist for each computer programming language inhabiting our real world.

In this chapter I have also provided evidence to support the historical claim that there never has been a standard model of computation throughout the history of computer science, even though I acknowledge that standard models do exist in specific quarters of computer science. For example, the lambda calculus plays a pivotal role in programming language design today and the Turing machine remains — for good or worse — central to the development of complexity theory. Nevertheless, it is the *quest* for an adequate model of computation and not the model itself that is the common denominator among computer scientists, including researchers with very similar backgrounds and research agendas.

To conclude, I address the question: Are proofs programs? A detailed study of the philosophical literature will hopefully convince the software scholar that proofs are not running computer programs [72, 233, 317]. The findings in the present chapter allow the scholar to go one step further: while proofs and mathematical programs belong to the same category, mathematical programs and computer programs belong to different categories. Hence, a proof is not a computer program, *irrespective* of whether the latter is running or not. So, stating that "proofs are programs" is only warranted with the proviso that we are referring to mathematical programs only.

7. TENACIOUS TALE about Functions and Relations: an Annotated Bibliography — by Raymond Boute

About the Author

Raymond Boute received MSc degrees in Electrical and Mechanical Engineering (1966) and in Electronics (1968) from Ghent University and a MSc degree (1969) and a Ph.D. degree (1973) in Electrical Engineering (with minor in Computer Science) from Stanford University. From 1973 to 1981 he worked at the Bell Telephone Mfg. Cy. (now Nokia), Antwerp, on the system architecture and implementation of computer-controlled telephone exchanges. In 1978 and 1980 he also spent two periods at the Intel research center in Aloha, Oregon, working on language design for systems programming of VLSI computers. From 1981 to 1994 he was a full professor of Computer Science at Nijmegen University (now Radboud University) teaching (successively) digital systems, machine architecture, computer networks, operating systems, declarative languages. He initiated the research on functional programming languages and declarative languages in Nijmegen. From 1994 until obligatory retirement age (2009) he was a full professor at Ghent University, doing teaching and research in programming language semantics and formal systems description, with particular emphasis on the common mathematical basis for classical engineering (analog models) and computer engineering (digital models, language semantics). His deepest interests have always been in electronics and radio engineering.

Abstract

Functions (and, by proxy, *relations*) are central concepts in mathematics. Their "modern" definition is very simple, and was free of issues 50 years ago, yet many recent books, often basic introductions to proofs, provide unsound definitions. An annotated bibliography is therefore a useful resource for students, instructors and authors. A reading companion exposes the pitfalls that require special caution, especially the notion of *codomain* which, for various technical reasons, turns out to be spurious. The most common errors appear to be caused by sloppiness and uncritical copying. A careful comparison of the references yields a valuable undergraduate-level object lesson demonstrating that definitions require thoughtful *design decisions*, with proper *justifications* on the part of their writers and critical scrutiny on the part of their readers by questioning understanding even for basic concepts. It also becomes clear that the misconceptions would have been nipped in the bud by the elementary use of symbolism supporting the textual definitions.

7.1   Introduction

Herstein calls the *function* (or *mapping*) *"probably the single most important and universal notion that runs through all of mathematics"* [175]. This observation refers to the "modern" concept of a function, which derives its generality and versatility from its simplicity. Of course, such a crucial concept deserves a flawless treatment. Formulations were indeed issue-free in nearly all textbooks 50 years ago, and still are in most basic analysis/calculus texts today. However, many other recent books, even those giving initiations to proofs, contain logical contradictions that cause unsoundness and fundamental misconceptions which significantly reduce simplicity and universality.

Hence an annotated bibliography can help authors, instructors and students to avoid all typical pitfalls. Here we present a random sample from diverse areas of mathematics including algebra, analysis/calculus, discrete math, logic, set theory etc. This diversity also invalidates the myth that different areas benefit from having different notions of a *function*.

7.2   Reading companion

This reading companion forms the backbone of the bibliography. It is not a dispensable luxury, since the simplicity and the presumed familiarity of the concepts tends to impair alertness.

The order of presentation is the same as in most texts listed: *relations* as an underlying concept, followed by *functions* as the concept of interest. Analysis/calculus texts usually skip relations. We focus on the *definition*, the property of being *from X to Y*, *equality*, *composition*, *inverses* and, for functions, also the meaning of $f : X \to Y$, *onto-ness* and the problematic *codomain*. The view on *equality* is the litmus test for understanding.

To avoid crosstalk, each subsection cites only texts using near-identical wording in defining a *relation* or a *function*. Reference numbers match the annotations in Section 7.3.

Convention   Clarity and uniformity in symbolic notation are served by respecting the distinction between the *use* and the *mention* of a symbolic name. In particular, by reserving the standard symbol "$\in$" (read as "is in" or similar) for set membership and adopting a suitable symbol such as ":" (read as "in") for name binding, this convention helps to make it clear that $x \in X$ ("x is in X") is a *statement* where x is *used*, and $x : X$ ("x in X") is a *declaration* where x is *mentioned* by way of introduction. The so-called *prose test* requires that a symbolic formula, transliterated symbol by symbol into natural language, should yield proper grammar. For instance, $S \cap T = \{s : S \mid s \in T\}$ is read as "The intersection of S and T equals the set of all s in S such that s is in T." The goal of this test is not so much readability (a fine bonus nevertheless) as obtaining a well-designed grammar for symbolic notation.

Of course, this convention is not guaranteed to apply in literal quotations.

7.2.1   Relations I: Simple and safe formulations

The simplest definition of a relation, referred to as "modern" 50 years ago, is typical in older texts [42, 312, 316] although it is less common today [194, 300, 348].

Definition 1 (Relation, domain, range) *A relation is a set of (ordered) pairs.*[106] Equivalently, in symbols [42, 312]:

$$R \text{ isrel} \quad \equiv \quad \forall z : R . \exists x . \exists y . (z = x, y) . \tag{7.1}$$

The domain *of a relation R is the set of the first members of the pairs in R. The* range *of a relation R is the set of the second members of the pairs in R.* Notation: the domain of R is written as $\mathcal{D} R$, $\text{dom}(R)$ or similar, and the range of R as $\mathcal{R} R$ or $\text{ran}(R)$ etc.

Clearly, $\{(0,2), (1,3)\}$ is a relation; its domain is $\{0,1\}$ and its range is $\{2,3\}$.

Views in the literature diverge on writing $x \, R \, y$ or $y \, R \, x$ for $(x, y) \in R$. Quine [282] gives several good reasons for following Peano and Gödel in writing "y is the father of x" as $y \, F \, x$. This can be reconciled with the usual order in coordinate pairs by letting $y \, R \, x$ stand for $(x, y) \in R$ and mentioning this under a conspicuous heading.

The following terminology is commonly used for classifying relations.

Definition 2 (Relation from X to Y) *A relation from X to Y is a relation whose domain is included in X and whose range is included in Y.* Equivalently, in symbols:

$$R \text{ isrel}(X, Y) \quad \equiv \quad R \text{ isrel} \wedge \mathcal{D} R \subseteq X \wedge \mathcal{R} R \subseteq Y . \tag{7.2}$$

Hence $\{(0,2), (1,3)\}$ is a relation from $\{0,1\}$ to $\{2,3\}$ but also from $\{0,1,7\}$ to $\{1,2,3\}$.

The bibliography further mentions the following important concepts.

Definition 3 (Composition) *The* composition *$S \circ R$ of relations S and R is the relation such that $z(S \circ R)x$ iff $z \, S \, y$ and $y \, R \, x$ for some y.* (Note: the composition of relations is sometimes called the *relative product* [312, 316] or *resultant* [282], and written S/R or $S|R$.)

Definition 4 (Inverse)
The inverse $R^{\smile}$ *of a relation R is the relation such that $x \, R^{\smile} \, y$ iff $y \, R \, x$.*

Definition 5 (Functional relation) [42, 244]
A relation R is functional *iff no two pairs in R have the same first member.*

7.2.2 Relations II: From combined to unsound formulations

Definitions 1 and 2 ensure safety through the *separation of concerns*, first defining *relation*, then *from X to Y*. With one exception [300], the *introduction to proof* books listed wrap both concepts together in the following way [36, 79, 80, 147, 148, 156, 164, 173, 290, 304, 330].

Definition 6 (Relation from X to Y: combined definition)
A relation from X to Y is a subset of $X \times Y$. Note: this can be written symbolically as

$$R \text{ isrel} (X, Y) \quad \equiv \quad R \subseteq X \times Y . \tag{7.3}$$

Clearly, $\{(0,2), (1,3)\}$ is a subset of $\{0,1\} \times \{2,3\}$ and of $\{0,1,7\} \times \{1,2,3\}$. More generally, proving equivalence between Definitions 2 and 6 is easy: RHS (7.2) $\equiv$ RHS (7.3).

Still, some texts reflect the misconception that Definition 6 attaches X and Y to the relation, and that one cannot even define just a *relation* without adding "*from X to Y*". Even when not stated explicitly, this misconception often becomes apparent in the view on *equality*, insofar as it is mentioned (otherwise, the interpretation remains open).

Relation equality was clear 50 years ago: sets are equal iff they contain the same elements, and a relation by Definition 6 is a set of pairs. Very few users of Definition 6 seem aware of this fact, and most "define" (in effect, *redefine*) equality, in some cases logically contradicting Definition 6. For instance, in [290] a relation R from X to Y and a relation S from U to V are equal only if $X = U$ and $Y = V$. In this manner, since $R := \{(0,2), (1,3)\}$ is a relation from $\{0,1\}$ to $\{2,3\}$ and from $\{0,1,7\}$ to $\{1,2,3\}$, equality of R to itself would require $\{0,1\} = \{0,1,7\}$ and $\{2,3\} = \{1,2,3\}$—a contradiction.

Many texts [290, 304, 330, 334] define $R \circ S$ only for a relation R from X to Y and a relation S from Y to Z, and some explicitly claim that, if S is from Y' to Z with $Y' \neq Y$, then $R \circ S$ is undefined. However, even if $Y' \neq Y$ we still have $R \subseteq X \times (Y \cup Y')$ and $S \subseteq (Y \cup Y') \times Z$, and thus $S \circ R$ is always defined even by the seemingly restrictive definition. The equality misconception causes the replacement of Y and Y' by their union to be seen as changing R and S whereas, in fact, these relations remain the same.

7.2.3   Functions I: Simple and safe formulations

As a tribute to simplicity and clarity, we quote Apostol [12]:

Definition 7 (Function) *A function f is a set of ordered pairs (x, y) no two of which have the same first member.* Example: $\{(0, 2), (1, 3)\}$ is a function.

Equivalent simple formulations appear in [42, 84, 140, 194, 242, 244, 300, 312, 316, 348], essentially following Bourbaki [42, p. 77] in defining a *function* as a functional relation. Functionality justifies writing $y = f(x)$ if $y\,f\,x$. Texts bypassing the concept of a *relation* introduce the domain $\mathcal{D}f$ (range $\mathcal{R}f$) of a function f as the set of first (second) members of the pairs in f.

A direct result is the equality theorem. Quoting Apostol [12],

Theorem 1 (Equality) *Functions f and g are equal if and only if (a) f and g have the same domain, and (b) $f(x) = g(x)$ for every x in the domain of f.*

Texts using Definition 7 classify functions as in the ISO standard [191, p. 15] by

Definition 8 (Function from X to Y) *A function from X (in)to Y is a function with domain X and range included in Y. Such a function is introduced by writing $f : X \to Y$.*

For Bartle [28], this is a *mapping from X to Y*. Definition 8 is used in [12, 84, 140, 194, 300, 348], the most lucid and complete explanation being given by Flett [140] and quoted in Section 7.3. Clearly $\{(0, 2), (1, 3)\}$ is a function from $\{0, 1\}$ to $\{2, 3\}$ but also from $\{0, 1\}$ to $\{2, 3, 7\}$.

A related concept is *onto-ness*, defined in [28, 120, 140, 172, 175, 194, 211, 242, 300, 316, 348] by

Definition 9 (Onto Y) *A function is onto Y, or surjective on Y, iff its range is Y.*

Linguistically, "onto" is a *preposition*, and appears as such in Definition 9. Similarly, in proper mathematical usage, being *onto* or *surjective* is not a predicate on functions, but a relation between functions and sets.

The notions of f being *into Y* ($\mathcal{R}f \subseteq Y$) or *onto Y* ($\mathcal{R}f = Y$) have duals, namely, f being *partial on X* ($\mathcal{D}f \subseteq X$) or *total on X* ($\mathcal{D}f = X$). Writing

$f : X \nrightarrow Y$ specifies that f is a *partial function from X to Y* [244], that is: $\mathcal{D} f \subseteq X$ and $\mathcal{R} f \subseteq Y$. Typically, functions in analysis are partial on $\mathbb{R}$ or $\mathbb{C}$, and the domain is then characterized separately.

For composition, most texts provide the following definition [12, 28, 140, 172, 242], or equivalent theorems based on Definition 3 [312, 316], depending on whether relations are skipped or inserted as an intermediate concept.

Definition 10 (Function composition) *The composition $g \circ f$ of functions g and f is the function such that (a) its domain consists of all x in $\mathcal{D} f$ for which $f(x) \in \mathcal{D} g$ and (b) for all x in this domain, $(g \circ f)(x) = g(f(x))$.*

Finally, most definitions for inverses [12, 28, 164, 172, 219, 242, 304, 307, 312, 316] amount to the following:

Definition 11 (Injectivity, function inverse) *A function f is* injective or 1-1 (one-to-one) *iff $f(x) = f(x')$ implies $x = x'$. The* inverse f^- *of a 1-1 function f is the function from $\mathcal{R} f$ to $\mathcal{D} f$ such that $f^-(f(x)) = x$ for all x in $\mathcal{D} f$.*

7.2.4 Functions II: Pure concept versus representation

Strictly speaking, the set of pairs is really a (set-theoretic) *representation* of a function, called its *graph*. This point is noted, for instance, by Halmos [172, p. 30], Lang [218, p. 38], Lee [224, p. 48], Royden [295, p. 8] and Spivey [305, p. 29].

A more conceptual view is traceable to Euler, assuming generalization to arbitrary domains. For instance, the ISO standard [191, p. 15] merely says that a *function* assigns to each value in its domain a unique value in its range. Here, the term "assigns" can be made precise by an *assertion* of the form $A(x, f(x))$, and $A(x, y)$ is called a *(functional) relation* by Bourbaki [42, pp. 16, 47]. The common elements can be captured as follows.

Definition 12 (Function, "pure" concept) *A function f is an entity fully specified by* (i) *a set, written as $\mathcal{D} f$, called the* domain *of f and* (ii) *a unique value, written as $f(x)$ or fx, for each x in $\mathcal{D} f$. Here* fully *means that $f = g$ iff* (i) *$\mathcal{D} f = \mathcal{D} g$ and* (ii) *$f(x) = g(x)$ for each x in $\mathcal{D} f$. The* range *of f, written as $\mathcal{R} f$, is defined to be the set $\{f x \mid x : \mathcal{D} f\}$.*

The conceptual equivalence to Definition 7 can be made explicit by defining the *graph of a function f*, written $\mathcal{G} f$, as being the set $\{(x, f x) \mid x :$

$\mathcal{D} f$} — clearly a functional relation. Composition and inverses for the conceptual function can be most conveniently defined via graphs, e.g., $\mathcal{G}\,(g \circ f) \triangleq \mathcal{G}\,g \circ \mathcal{G}f$ and $\mathcal{G}f^{-} \triangleq (\mathcal{G}f)^{\smile}$, provided that $(\mathcal{G}f)^{\smile}$ is a function (injectivity of f). At least one textbook combines abstraction and simplicity in this manner; however, to my regret, I have lost the reference.

The equivalence between Definitions 7 and 12 can also be left implicit as just a change of perspective, as explained in [172, 295].

In fact, Halmos remarks in various places [172, pp. 25, 45] that identifying concepts with sets causes "freak properties", such as $(x, f\,x) \in f$, which can be either ignored, removed by axiomatization, or hidden by encapsulation. The latter approach is similar to data abstraction in computer science and, for functions, amounts to the inverse operator $\mathcal{G}'$ of $\mathcal{G}$. An abstract data type has at most as many attributes as its representation, but nearly always fewer: in particular, the "freak properties" are "abstracted away".

The symbols $\mathcal{D}$, $\mathcal{R}$, $\mathcal{G}$ used thus far require some comment. They are *functional* in the sense that the expressions $\mathcal{D}f$, $\mathcal{R}f$, $\mathcal{G}f$ are *unambiguous*, i.e., $f = g$ implies $\mathcal{D}f = \mathcal{D}g$ and so on.

More generally, non-ambiguity is captured by *Leibniz's rule* as formulated by Gries and Schneider [164], and very briefly summarized here. Let d and e be expressions and x a variable, then $d[^x_e$ denotes the expression obtained by properly substituting e for every occurrence of x in d. Given expressions e and e' and an assignment of values to the variables, the equality $e = e'$ indicates that evaluating e and e' for that assignment yields the same value. Leibniz's rule requires that, if $e = e'$, then $d[^x_e = d[^x_{e'}$. For instance, let d be $\mathcal{D}x$, and let e be f and e' be g; then $f = g$ implies $\mathcal{D}f = \mathcal{D}g$. Clearly, Leibniz's rule, functionality and non-ambiguity are equivalent concepts.

Although $\mathcal{D}$, $\mathcal{R}$, $\mathcal{G}$ are functional in the sense of Leibniz's rule, in elementary set theory they are not functions since their domain would be too "large" to avoid inconsistencies. In Lamport's terminology [216], they would be called *operators*. In variants of set theory that safely support the aforementioned large sets, often called *classes*, these operators can be considered functions again [316].

Another comment: saying that an expression e (possibly an atomic symbol, like cos) "is" a function is a shorthand form of saying that e is an expression that denotes a function. Such distinctions are usually clear from the context, making the shorthand form safe. Similarly, a phrase such as "Two functions f and g are equal iff ...", when taken strictly, is nonsensical: talking about *two* functions requires them to be different. Hence a proper phrasing would be: "Two expressions f and g denote the same function iff ...". The reader should also be warned that some philosophical texts use the term *name* (in normal usage atomic) for an *expression* (not necessarily atomic). The terminology in [164] is more appropriate.

7.2.5   Functions III: Convoluted formulations

The simplicity of Definitions 7 and 8 is often undervalued in favor of involving the Cartesian product (despite well-known educational reservations [302]) and wrapping both definitions together. This is still achieved safely in [28, 148, 172, 175, 213, 330], but not in [36, 62, 79, 80, 147, 149, 173, 290, 304, 334] due to an unsound addition, discussed later.

Definition 13 (Function from X to Y) *Let X and Y be sets.*

(a) *A function f from X to Y, written $f : X \to Y$, is a relation $f \subseteq X \times Y$ such that for each x in X the relation f contains exactly one ordered pair of the form (x, y).*

(b) *The set X is called the* domain of f.

As a warning, we have faithfully reproduced the nearly uniform but unacceptable phrasing which suggests that the function is written $f : X \to Y$ (in fact, the function is written just f), and that $f \subseteq X \times Y$ is a relation (in fact, it is a statement about the relation f).

More crucial is soundness. Few seem to realize that Definition 13(b) carries a proof obligation, namely, showing that X is indeed fully determined by f. This is simple, and reveals X to be the set $\mathcal{D} f$.

In fact, Definition 13 is equivalent to Definition 7, only rather convoluted. The equivalence was evident to everyone fifty years ago, but is obscured today by various misconceptions, rooted in the false belief that Definition 13 attaches Y to f. An effective remedy is a simple analysis of Definition 13 (exercise). The results are summarized as follows.

Lemma 2 (Dissecting Definition 13) *f is a function from X to Y according to Definition 13 iff f is a functional set of ordered pairs, $\mathcal{D} f = X$ and $\mathcal{R} f \subseteq Y$.*

Theorem 3 (Equivalence of Definitions 7/8 and 13) (I) *A function f as in Definition 7 is a* function from $\mathcal{D} f$ to $\mathcal{R} f$ *(or any superset of $\mathcal{R} f$) as in*

Definition 13. Conversely, a function from X to Y *as in Definition 13 is a* function *as in Definition 7.*

(II) *Definitions 8 and 13 define exactly the same concept of a* function from X to Y.

The differences are related only to formulation. However, they strongly affect clarity. The design of Definition 13 is poor, since it has given rise to absurd myths, Myth # 1 being that one cannot define the concept of a *function* by itself, but only *a function from X to Y*. Evidently, this myth is invalidated in [12, 42, 84, 140, 194, 242, 244, 300, 312, 316, 348].

Among the texts using Definition 13, very few classical ones [175, 295] define composition $g \circ f$ only for functions $f : X \to Y$ and $g : Y \to Z$, whereas nearly all current texts succumb to this restriction [36, 290, 300, 304, 330]. This is truly restrictive. Indeed, the earlier reasoning for $S \circ R$ does not apply to functions because, given $f : X \to Y$ and $g : Y' \to Z$ with $Y' \neq Y$, it is always true that $f \in X \to (\mathcal{R}f \cup \mathcal{D}g)$, but $g \in (\mathcal{R}f \cup \mathcal{D}g) \to Z$ holds only if $\mathcal{D}g = \mathcal{R}f \cup \mathcal{D}g$ (domain equality) or, equivalently, $\mathcal{R}f \subseteq \mathcal{D}g$. The requirement $\mathcal{R}f \subseteq \mathcal{D}g$ appears in [79, 80, 194, 213]. Still, calculus texts use the general form [219, 307], since this is necessary for practical applications.

7.2.6   Functions IV: Unsound formulations in recent texts

In texts published after 2000 where Definition 13 is adopted, just the definition of *onto-ness* has already become problematic. Some authors [148, 304] still properly write "onto Y", albeit with an explanation that lacks generality. Many others [36, 79, 80, 173, 213, 290, 330] use "onto" as an adjective, defining a function $f : X \to Y$ as being *onto* or *surjective* iff $\mathcal{R}f = Y$. This definition is unsound because it allows a function f to be both *onto* and *not onto* depending on whether or not the set Y chosen to appear in $f : X \to Y$ is $\mathcal{R}f$.

Furthermore, all *transition to proof* texts in our random sample [36, 62, 79, 80, 147, 149, 173, 290, 304, 334] except [148], make the following addition to Definition 13.

Definition 14 (Codomain) Definition 13 plus: (c) Y *is called the* codomain of f.

Like Definition 13(b), adding (c) carries a proof obligation: showing that Y is determined by f. This reveals a fatal contradiction: since

$\{(0,2),(1,3)\}$ is a function from $\{0,1\}$ to $\{2,3\}$ and from $\{0,1\}$ to $\{2,3,7\}$, its codomain equals both $\{2,3\}$ and $\{2,3,7\}$.

The unsoundness may be even clearer in symbolic form. Most texts complete (c) with a phrase such as "written $\mathcal{C}f$" (or cod (f) or similar) and use the convention that declaring $f : X \to Y$ specifies $\mathcal{C}f = Y$. Leibniz's rule is violated in the sense that $f : X \to Y$ and $g : X \to Y'$ may satisfy $f = g$ by Definition 13 although $\mathcal{C}f \neq \mathcal{C}g$, namely in the case of $Y \neq Y'$.

An often-heard excuse is that Definition 14 is just a figure of speech; however, most cited texts mean it literally. Indeed, the views on function equality are most revealing. Apart from [79, 148, 304], all *transition to proof* texts using Definition 14 seem unaware of Theorem 1 and define equality anew. In the best case, part (c) is ignored [79, 148, 290, 304], even intentionally [304]. In the worst case, Definition 13 is contradicted by requiring equal codomains [36, 147, 173]. Some texts [62, 80] escape contradiction by defining equality only for the case when both functions are from X to Y, but this is too restrictive in practice.

7.2.7 Technical evaluation: Universality versus codomains

Repairing Definition 14 is easy, provided one recognizes its incompatibility with Definition 13 and accepts that it imposes a *conceptually different* notion of a *function*. For instance, as in Bourbaki [42, p. 76],

Definition 15 *A function f from X to Y is a triple (F, X, Y) where F is a functional relation satisfying $F \subseteq X \times Y$ and $\mathcal{D}F = X$.*

Ironically, on the very next page [42, p. 77], Bourbaki announces the use of *"function"* for just a functional set of pairs, as in Definition 7! This explains Halmos's amusement [171] in observing that the Bourbaki group often abandon their "innovations" in favor of common terminology. Arguably, the triples definition is just an unfortunate accident [42, 84].

Indeed, apparently more difficult than repair is exercising adequate judgment in the choice of definitions. McCabe's observation that "any clod can have the facts, but having opinions is an art" can be directly paraphrased here.

In fact, definitions and axiomatizations are not arbitrary starting points for a game of logic. Rather, they reflect *design decisions* that require justification and evaluation [45]. The main criteria directly reflect the very purpose of a definition/axiomatization.

One criterion is *conceptualization*: how well does a definition/axiomatization capture the concept of interest? Simplicity and universality are the hallmarks of good design.

Another criterion is *practicality*: how well does a definition/axiomatization serve mathematics at large and applications both in and outside of mathematics?

Clearly such criteria are not independent. Moreover, few mathematicians justify their design choices, a laudable exception being Halmos [172].

As regards codomains, the only evaluation found in the literature is by Shuard [302]. Shuard notes that Definition 7 as stated by Flett [140] "wins hands down" as far as simplicity in analysis is concerned, but suggests two merits for attaching codomains in algebra: (i) using "onto" as an adjective, and (ii) writing $f : X \to Y$ to *mention Y*. However, these merits are delusive since (i) the ability to say "f is onto Y but not onto Z" is more selective, and (ii) under the standard interpretation, $f : X \to Y$ mentions Y equally well, which is also demonstrated amply in algebra [175].

Still, one might ask whether there is any harm in attaching a codomain to a function. In fact, the harm is considerable, as shown by the following additional technical points. Each point briefly recalls a conceptually and practically valuable property of functions (in the standard sense) that would be impaired, often fatally, by attaching codomains. Verifying the details is left as a simple exercise.

(a) In the standard interpretation, $f : X \to Y$ specifies that $\mathcal{D} f = X$ and $fx \in Y$ for all x in $\mathcal{D}f$. Hence the expression $X \to Y$, called a *function type*, can serve as a *partial specification* for f. This provides significant flexibility for specifying functions in the style that is most expressive for the reader and the writer. For instance, the declarations

$$\text{sqrt} : \mathbb{R}_{\geq 0} \to \mathbb{R}_{\geq 0} \quad \text{with} \quad (\text{sqrt } x)^2 = x$$
$$\text{sqrt} : \mathbb{R}_{\geq 0} \to \mathbb{R} \quad \text{with} \quad (\text{sqrt } x)^2 = x \text{ and sqrt } x \geq 0$$

specify exactly the same function, whose graph is $\{(x,y) : \mathbb{R}_{\geq 0} \times \mathbb{R}_{\geq 0} \mid y^2 = x\}$. Attaching the set Y mentioned in $X \to Y$ to f as a codomain forfeits such opportunities.

(b) By Definition 10, the composition $g \circ f$ is defined for *any* functions f and g, and is associative. In texts using codomains, $g \circ f$ is defined only for cases where $\mathcal{D}g = \mathcal{C}f$. This is unacceptably restrictive for practical areas of mathematics, especially calculus.

(c) By Definition 11, any 1-1 function f has an inverse f^- with domain $\mathcal{R}f$. If one attaches a codomain, a 1-1 function from X to Y has an inverse only if it is "onto". Shuard explains how this causes errors in the only known calculus text that uses codomains.

(d) The term *family* is a synonym for *function* in many areas of science (physics, engineering) and in mathematics [42, 172]. For instance, a *family of sets* is just a set-valued function. Attaching a codomain would be awkward.

(e) Writing $f : X \to Y$ specifies the uniform constraint $\forall x : X . f(x) \in Y$. More general and selective is a constraint of the form $\forall x : X . f(x) \in T(x)$. This idea is captured by the *Cartesian product* operator $\times$ as follows [172]: for any family T of sets, $\times T$ is the set of functions f satisfying $\mathcal{D}f = \mathcal{D}T$ and $\forall x : \mathcal{D}f . f(x) \in T(x)$.

(f) An *n-tuple* or *sequence* of length n is a function[107] on the set of the first n natural numbers [172, 213, 295, 296], henceforth written $\square n$. For instance, the triple $(a, 27, c)$ denotes a function with domain $\square 3$. Attaching a codomain would require awkward bookkeeping.

Letting $T := (X, Y)$, which has domain $\square 2$, and switching to infix notation by writing $\times (X, Y)$ as $X \times Y$ yields the familiar Cartesian product (exercise). This formulation also resolves the "terminological friction" discussed by Halmos [172].

Similarly, $X \times Y \times Z \triangleq \times (X, Y, Z)$ and so on. Clearly $(X \times Y) \times Z \neq X \times Y \times Z$. Thus one can cover tree structures, so crucial in computer science. This opportunity is forfeited if (x, y, z) is identified with $((x, y), z)$ and $X \times Y \times Z$ with $(X \times Y) \times Z$, as in Bourbaki [42, p. 70]. The same holds for matrices.

(g) Matrices, trees and similar structures are covered in the same uniform way. An n by m matrix can be defined over a set Z as a function $M : \square n \times \square m \to Z$ and an entry in row i and column j can be written as $M_{i,j}$, without requiring the wasteful mix of upper- and lowercase symbols. One can equally well define $M' : \square n \to (\square m \to Z)$ such that $(M'i)j = M_{i,j}$. This shift of perspective, known as *currying* [34, 244, 273], can be generalized in the simplest case to mapping a function $f : X \times Y \to Z$ to a function $g : X \to (Y \to Z)$ with $(g\,x)y = f(x, y)$. The *currying* operator ^{C} allows g to be written as f^C and has an inverse $^\supset$. Similarly, *transposition* maps $g : X \to Y \to Z$ to $h : Y \to X \to Z$ with $h\,y\,x = g\,x\,y$.

The above points are only a small sample; more can be found in [43, 244].

Most fallacies about codomains in the literature reflect an inadequate understanding of the standard interpretation of $f : X \to Y$, in particular the fact that there is no need to attach Y to f to obtain the effects that Shuard considers desirable in algebra.

An inescapable conclusion is that the concept of codomain, as defined in recent texts, is spurious since it is either unsound or, if repaired (by redefining *function*), harmful in mathematics as a whole, as it destroys the universality and power of the function concept.

7.3 Annotated bibliographical list

Items written *"thus"* are literal quotes, using an ellipsis (. . .) for parts that are evident (free of issues), and respecting punctuation and spaces in formulas as much as possible, occasionally revealing unpalatable syntax. Items written [thus] are slightly shortened but otherwise faithful quotes. The remainder are annotations, often using "telegraph style". Page numbers are included to encourage actually looking up the information, since even literal quotes are no substitute for the full text.

[12] Apostol, *Calculus*

- *Function*: page 53. "DEFINITION. *A function f is a set of ordered pairs (x, y) no two of which have the same first member.*"
  Page 54. [THEOREM 1.1. $f = g$ iff dom f = dom g and $f(x) = g(x)$ for all x in dom f.] Note: few texts today seem aware of this.
  Page 140. Composition, fully general. If $f = u \circ v$, "*f will be defined only for those points x for which $v(x)$ is in the domain of u.*"
  Page 146. Inverse, fully general. The inverse of a 1-1 function f has domain $\mathcal{R} f$ and range $\mathcal{D} f$.
  Page 578. "*The symbol $T : V \to W$ will be used to indicate that T is a function whose domain is V and whose values are in W.*"

[28] Bartle, *The elements of real analysis*

- *Function*: page 13. "*Let A and B be sets. A **function from A to** B is a set f of ordered pairs in $A \times B$ with the property that if (a, b) and (a, b') are elements of f, then $b = b'$.*"
  Note: here A may be a superset of the domain of f. However:

Page 13. [Notation $f : A \rightarrow B$ to indicate that f is a *mapping* from A to B, i.e., a function from A to B satisfying $\mathcal{D}(f) = A$.]

Page 13. "*If $\mathcal{R}(f) = B$, we say that f maps A **onto** B.*" Observe the proper use of "onto", as a preposition.

Page 16. Defines $g \circ f$ as for relations; $\mathcal{D}(g \circ f) = \{x : \mathcal{D}f \mid fx \in \mathcal{D}g\}$ is a theorem.

Page 17. Defines f as being 1-1 iff $(a, b) \in f$ and $(a', b) \in f$, then $a = a'$.

Page 18. Defines the inverse of a 1-1 function f rather clumsily, mixing elements of a theorem and a definition together, but essentially as $\{(b, a) : \mathcal{R}f \times \mathcal{D}f \mid (a, b) \in f\}$.

[36] Bloch, *Proofs and Fundamentals*

- *Function*: page 131. "**Definition 4.1.1.** *Let A and B be sets. A **function** (or **map**) f from A to B, denoted $f : A \rightarrow B$, is a subset $F \subseteq A \times B$ such that .... The set A is called the **domain** of f and the set B is called the **codomain** of f.*"
  The ambiguous switch from f to F is patched as if by afterthought:
  Page 131. "*One way of defining a function is as a triple (A, B, F)*".
  Due warnings against this interpretation are found under item [84].
  Page 136. Function equality is redefined as requiring equal codomains.
  Page 146, Def. 4.3.1. Composition only for $f : A \rightarrow B$, $g : B \rightarrow C$.
  Page 149, Def. 4.3.6.3. Inverse of $f : A \rightarrow B$ is from B to A, which requires $B = \mathcal{R}f$.
  Page 155, Def. 4.4.2. Onto is used as an adjective (assumes codomain).
- *Relation*: page 172. "**Definition 5.1.1.** *Let A and B be sets. A **relation** R from A to B is a subset $\overline{R} \subseteq A \times B$. If $a \in A$ and $b \in B$, we write a R b if $(a, b) \in R$ and a $\mathcal{R}$ B if $(a, b) \notin \overline{R}$.*"
  Note: again, the switch between R and $\overline{R}$ is carefully reproduced.
  Page 173. "**Definition 5.1.3.** *The **relation class** of x with respect to R, written R[x], is the set defined by $R[x] = \{y \in B \mid x R y\}$.*"
  Note: this illustrates why $y R x$ is preferable over $x R y$; see [282].

[42] Bourbaki, *Théorie des ensembles*. Here a brief legend is required. Bourbaki uses the term *relation* (p. 16) for an expression R denoting an *assertion* and calls R *functional w.r.t. x* (p. 48) if exactly one x satisfies R. Present-day terms for Bourbaki's *relation* are *assertion* and *statement*. In these annotations, the term *relation* is used in its present-day sense (set of ordered pairs), as everywhere else in this bibliography.

- *Relation*: page 71 (called *graphe*). "*On dit que G est un graphe si tout élément de G est un couple.*" Note: *couple* (p. 68) means *ordered pair*, not the concept from physics.

  Page 72. "*On appelle correspondance entre un ensemble A et un ensemble B un triplet* $\Gamma = (G, A, B)$ *où G est un graphe tel que* $\mathrm{pr}_1 G \subseteq A$ *et* $\mathrm{pr}_2 G \subseteq B$."

  Page 73. The inverse relation (*graphe réciproque*) $\overset{-1}{G}$ of G is $\{(y, x) \mid (x, y) : G\}$.

  Page 73. The inverse $\overset{-1}{\Gamma}$ of $\Gamma := (G, A, B)$ is the correspondence $(\overset{-1}{G}, B, A)$.

  Page 74. The composite $G' \circ G$ of G and G' is $\{(x, z) \mid \exists y \,.\, (x, y) \in G \wedge (y, z) \in G'\}$.

  Page 75. The composite $\Gamma' \circ \Gamma$ of $\Gamma := (G, A, B)$ and $\Gamma' := (G', B, C)$ is $(G' \circ G, A, C)$.

  Observe how the ballast makes correspondences more restrictive than relations.

  Page 76. "*On dit qu'un graphe F est un graphe fonctionnel si, pour tout x, il existe au plus un objet correspondant à x par F.*"

- *Function*: page 76. First variant: "*On dit qu'une correspondance f = (F, A, B) est une fonction si son graphe F est un graphe fonctionnel et si* $A = \mathrm{pr}_1 G$."

  Page 76. "*On appelle* application *de A dans B une fonction dont l'ensemble de départ est égal à A et dont l'ensemble d'arrivée est égal à B.*" **However:**

  Page 77. Second variant: "*Nous employerons souvent, dans la suite de ce Traité, le mot "fonction" à la place de "graphe fonctionnel".*" and (p. 77) "*Dans certains cas, un graphe fonctionnel s'appelle aussi une famille.*", as in Halmos [172, p. 34].

  Clearly Bourbaki slips back into the simpler common conventions, as often happens, to Halmos' amusement [171, p. 80]. This is a wise decision, since it bestows on function inverse and composition the most general form, as relation inverse and composition.

  Bourbaki avoids confusion between a *function* as a functional relation and the triple-based variant by calling the latter an *application*. He defines *injective*, *surjective* and *bijective* for applications only, although "*injective*" is meaningful for graphs as well.

  Page 80. "*Soit f une application de A dans B. On dit que f est une injection, ou que f est une application injective, si deux éléments distincts de A ont des images distinctes par f. On dit que f est une surjection, ou que f est une application surjective, si f(A) = B. On dit que f est une bijection, ou que f est une application bijective, si f est à la fois injective et surjective."* Obviously, applying the terms *surjective* and *bijective* to a functional relation requires explicitly mentioning a target set.

[79] Daepp & Gorkin, *Reading, Writing and Proving*

- *Relation*: page 101. "*A **relation from** X **to** Y is a subset of X × Y.*"
- *Function*: page 147. "*A **function** f from A to B is a relation from A to B satisfying . . . . We usually write f : A → B to indicate that f is a function from A to B.*"

  Page 148. Codomain as a function attribute (conflict with the definition).

  Page 152. Equality criterion derived from the definition, essentially Theorem 1.

  Page 163. *Onto* is used as an adjective (assumes codomain).

  Page 175. Composition $g \circ f$ of $f : A \to B$ and $g : C \to D$ defined in case $\operatorname{ran}(f) \subseteq C$.

  Page 177. Inverse of 1-1 function $f : A \to B$ is from B to A, which requires $B = \mathcal{R} f$.

[80] Second edition of [79], with only one relevant change:

Page 146. Equality discussed only for $f : A \to B$ and $g : A \to B$.

[84] Dasgupta, *Set Theory*

- *Relation*: page 8. A relation *is defined to be any set consisting only of ordered pairs. Thus:*

  R *is a relation* $\Leftrightarrow$ *for all* $x \in R$, $x = \langle a, b \rangle$ *for some* a, b.

  Page 8. Domain and range are defined as usual.
- *Function*: page 10. "*A relation F is said to be a* function *if*

  $x \, F \, y$ *and* $x \, F \, z \Rightarrow y = z$ *for all* x, y, z."

  Page 10. "*We also say that F is a function from A to B, written using the standard notation F : A → B, to mean that F is a function with* $\operatorname{dom}(F) = A$ *and* $\operatorname{ran}(F) \subseteq B$. *In this case, it is common to abuse terminology and refer to the triplet* $\langle F, A, B \rangle$ *as "the function F : A → B." The set B is then sometimes referred to as the* converse domain *or co-domain of the function F : A → B (more precisely, B is the co-domain of the triplet* $\langle F, A, B \rangle$)."

  Note: "abuse terminology" is a euphemism for "mix up concepts". Dasgupta clarifies the distinction between the function F and a triplet like $\langle F, A, B \rangle$: only triplets have codomains. This distinction is also important in understanding the relationship between *relations* and *functions* in their ordinary mathematical meaning and *arrows* in category theory, which captures the corresponding triplets. This is explained very well in [34, p. 26] and in [273, p, 2].

[120] Exner, *An Accompaniment to Higher Mathematics* Note: right from the start, this text recalls certain concepts known from earlier experience, and (re)defines them later on, as reflected by some double definitions below. Exner's style of exposition with continuous questioning might have been a good opportunity for stimulating critical thinking but, as we shall see, was not exploited thoroughly enough.

- *Relation*: page 5, 135 "*A relation from A to B is a subset of A × B.*"
- *Function*: page 5 "**Definition**   *A function with range contained in S is surjective on S if for every s in S there is an x in the domain of f such that f(x) = s.*" This is a roundabout way to say that the range of *f* is *S*, but the crucial point is the phrasing *surjective on S* rather than just *surjective*.

  Page 16. "*When we discuss a function, say f, it is to be understood that we have in mind a certain domain [denoted domain(f)] and a certain codomain [denoted codomain(f)]. By codomain(f) we mean some fixed set such that f(x) ∈ codomain(f) for each x in domain(f).*" Observe the usual unsoundness w.r.t. the following.

  Page 20. "**Definition**   *A function from A to B is a relation from A to B such that if (a, b) and (a, c) are in f then b = c.*" On the other hand, observe the cautious phrasings, such as "*Let f be a function with domain X and Y some codomain*" (rather than "*... and codomain Y*") in items **1.63** and **1.64** (same page). Clearly the author is uneasy about attaching *Y* to *f*. Unfortunately, all caution is cast aside on p. 123.

  Page 123. "**Definition 4.2.1**   *Given sets A and B, a function from A to B is a subset of A × B (...) with the property that each element a ∈ A occurs as the first member of an ordered pair exactly once (...). The set A will be called the domain of f and denoted domain(f), and the set B the codomain of f and denoted codomain(f).*" This opens the possibility that *f* = *g* although *codomain(f)* ≠ *codomain(g)*, causing unsoundness by violating Leibniz's principle.

  Page 124. "**Definition 4.2.2**   *Given a function f from domain(f) to codomain(f), the range of f (denoted range(f)) is the subset of codomain(f) consisting of those elements that occur as the second element of some ordered pair in f.*" Here the design principle of *separation of concerns* is violated by involving codomains in a situation where they make no difference. Indeed, why not simply: "The range of a function f (denoted $range(f)$) is the set consisting of those elements that occur as the second element of some ordered pair in f."? Caution partially resurfaces in the following:

Page 127. "**Definition 4.2.5** *A function f with codomain B is surjective on B (alternatively, onto B) if each element of B occurs as the second element of an ordered pair of B at least once. If the codomain B is understood, we sometimes simply say that f is surjective or onto or that f is a surjection.*" Apart from the roundabout phrasing as on page 5 of [120], observe that the phrase "*'with codomain B*" is redundant (violating separation of concerns) and reduces generality. Moreover, the phrase "*If the codomain B is understood*" clearly implies the possibility that the codomain is *not* an attribute of the function, which is similar to Smith's view [304].

Page 127. **Definition 4.2.7** defines composition $g \circ f$ only if $range(f) \subseteq domain(g)$. This is restrictive, but less restrictive than requiring $codomain(f) = domain(g)$. Exner also notes that defining composition and inverses (page 128) via graphs makes things easier.

Page 128. Exner lets the reader develop the concept of inverse function in successive steps. The statement in item **4.29**, "*Recall that, intuitively, the inverse of a function from A to B is some function from B to A [...]*" reduces generality by requiring the function to be onto B. This might well be a glitch, since item **4.31** invites the reader to verify that $f \circ f^{-1}$ is the identity function on $range(f)$, which is fully general for 1-1 functions.

[140] Flett, *Mathematical Analysis*

* *Function*: page 4 "*A function is a set of pairs with the property if $(x, y) \in f$ and $(x, y') \in f$, then $y = y'$.*"
The set of first/second coordinates of the pairs in f are called the *domain/range* of f.

Page 5. "*For any sets X and Y, a function f with domain X and range contained in Y is called a* function from X into Y; *if the range is identical to Y, we say also that f is a* function from X onto Y. *It must be stressed that we use these two phrases merely to convey information about the domain and range of f, and that we are not here defining a new mathematical object. Thus, if Y and Y' are two sets both containing the range of f, then "the function f from X into Y" and "the function f from X into Y'" both refer to the same object, namely the function f, i.e., the set of ordered pairs.*"

Page 6. "*It follows immediately from the definitions that, if f_1 and f_2 are functions, then $f_1 = f_2$ if and only if f_1 and f_2 have the same domain and for every element x of this common domain we have $f_1(x) = f_2(x)$.*"

(Observe that this is a theorem and not a definition, since the statement $f_1 = f_2$ means that the two sets of ordered pairs f_1 and f_2 are identical.)" If only more texts reflected this insight!

Page 6. A footnote mentions defining a "composite object" (f, X, Y) as an alternative that *"is sometimes desirable in certain parts of mathematics"*. However, the literature provides no justification for any such desirability that, on closer analysis, does not turn out to be fallacious (e.g., [302]), typically overlooking the fact that $f : X \rightarrow Y$ already specifies that $\mathcal{D}(f) = X$ and $\mathcal{R}(f) \subseteq Y$, which suffices for all sensible purposes.

Page 8. An interesting aside: $f(A)$ is defined for *any* set A, regardless of $\mathcal{D}(f)$.

Page 9. Similarly, $f^{-1}(B)$ is defined for *any* set B, regardless of $\mathcal{R}(f)$.

Page 11. Composite: $\mathcal{D}(f \circ g) = \{x : \mathcal{D}(g) \,|\, g(x) \in \mathcal{D}(f)\}$ and $(f \circ g)(x) = f(g(x))$.

Page 12. For 1-1 f, $f^{-1} = \{(y, x) \,|\, (x, y) : f\}$, hence $\mathcal{D}(f^{-1}) = \mathcal{R}(f)$, $\mathcal{R}(f^{-1}) = \mathcal{D}(f)$.

[147] Garnier & Taylor, *Discrete Mathematics — Proofs, Structures and Applications*

- *Relation*: page 155. "**Definition 4.1** Let A and B be sets. A **relation from A to B** is a subset of $A \times B$."

  Page 173. Defines $S \circ R$ for R from A to B and S from B to C.

- *Function*: page 224. "**Definition 5.1** A **function** f **from A to B**, written $f : A \rightarrow B$, is a relation from A to B which . . . ."

  Page 224. Attaches a codomain (conflicts with the definition).

  Page 224, Def. 5.2. Equality is redefined: given $f : A \rightarrow B$ and $g : A' \rightarrow B'$, $f = g$ also requires $B = B'$. A justification is promised on p. 225, but not found. An email from Taylor suggests a possibility:

  Page 247. *Surjective function* is defined without an explicit set.

  Page 238. Composition $g \circ f$ only for $f : A \rightarrow B$ and $g : B \rightarrow C$ (restrictive).

  Page 264. Inverse of injective $f : A \rightarrow B$ defined assuming $B = \mathcal{R} f$.

  Page 138. An interesting aside: at this stage, the text introduces *typed set theory*, which occasionally resurfaces in the book. The motivation is software, and the very restricted view on types (every object having a unique type) is a far-going concession to the limitations of current technology for type algorithms. In the mechanization of mathematics, similar implementability issues arise. Yet, conceptually, as in "ordinary mathematics", this is quite stifling. A very enlightening discussion is found in [217].

[148] Gerstein, *Introduction to Mathematical Structures and Proofs*

- *Relation*: page 86. "**2.50 Definition.** *A **relation** R on a set S is a collection of ordered pairs of elements of S, i.e, R $\subseteq$ S $\times$ S.*"
  Page 96, Exercise 19. "*Many books define a* relation *from A to B to be a subset R $\subseteq$ A $\times$ B.*"

- *Function*: page 110. "**3.2 Definition.** *A **function** f from A to B is a set of ordered pairs f $\subseteq$ A $\times$ B with the property . . . . We usually write f : A $\to$ B to indicate that f is a function from A to B.*"
  Page 111. Careful mention of *codomain* as a *context* attribute.
  Page 113, item (3.4). Equality correctly derived from the definition.
  Page 118. "*Given sets A and B, a function f : A $\to$ B is said to be **surjective** or **onto** B if f(A) = B, and f is a **surjection***"
  Note: "onto B" is defined correctly but lacking generality (f : A $\to$ C can also be *onto B*); "surjective" is unsound without mentioning a target set.
  Page 131. Composition $g \circ f$ only for $f : A \to B$ and $g : B \to C$ (restrictive).
  Page 135. Inverse f^{-1} of injective $f : A \to B$ defined assuming $B = \mathcal{R} f$.

[156] Goodaire & Parmenter, *Discrete Mathematics with Graph Theory*

- *Relation*: page 51. "**2.3.1 Definitions.** *Let A and B denote sets. A* binary relation *from A to B is a subset of A $\times$ B.*"

- *Function*: page 72. "**3.1.1 Definition.** *A **function** f from a set A to a set B is a binary relation with the property that . . . .*"
  Page 73. "**3.1.2 Function notation** *It is customary to write f : A $\to$ B to mean that f is a function from A to B.*"
  Page 73, Def. 3.1.3. *B* is defined as the *target* of *f* (conflict). *Onto* is defined as an adjective, requiring the target.
  Page 80, Def. 3.2.1. The inverse of $f : A \to B$ is considered to exist iff $\{(b, a) \mid (a, b) : f\}$ is a function from *B* to *A* (restrictive).
  Page 82, Def. 3.2.4. Composition $g \circ f$ only for $f : A \to B$ and $g : B \to C$ (restrictive).
  Page 83, Def. 3.2.5. Equality redefined, requiring the same target.

[164] Gries & Schneider, *A Logical Approach to Discrete Math*

- *Relation*: page 267. A relation over $B \times C$ is defined as a subset of $B \times C$.

Page 270. Inverse is defined as usual: $\langle c, b \rangle \in \rho^{-1} \equiv \langle b, c \rangle \in \rho$ (misprint corrected).

Page 271. The *product* $\rho \circ \sigma$ of a relation ρ on $B \times C$ and a relation σ on $C \times D$ is defined by $\langle b, d \rangle \in \rho \circ \sigma \equiv \{\exists c \mid c \in C : \langle b, c \rangle \in \rho \wedge \langle c, d \rangle \in \sigma\}$ (in the notation of [164]). This is restrictive only in appearance: if σ is on $C' \times D$ replace C and C' by $C \cup C'$.

- *Function*: page 280. "*(14.37)* **Definition.** *A binary relation f on $B \times C$ is called a function iff it is* determinate." Note: *determinate* means functional. Furthermore: "*A function f on $B \times C$ is* total *iff Dom.f = B.*" Note: clearly this should be *total on B*, since f itself does not fully determine B.

Page 281. Composition $f \bullet g = g \circ f$. See Quine's comments [282] on this notation.

Page 282. The function inverse is the relational inverse (provided it is a function).

Page 282. "*(14.41)* **Definition.** *Total function $f : B \to C$ is* onto *or* surjective *if Ran.f = C.*" Clearly, this should be *onto C* for soundness and selectivity.

The text also states (same page): "*A function can be made* onto *by changing its type.*"; however, the function remains the same set of pairs, and the type is not unique!

[172] Halmos, *Naive Set Theory*

- *Relation*: page 26. "*A set R is a relation if each element in R is an ordered pair.*"

Page 27. "*If R is a relation included in the Cartesian product $X \times Y$, it is sometimes convenient to say that R is a relation from X to Y.*"

- *Function*: page 30. "*A function from X (in)to Y is a relation f such that* dom $f = X$ *and such that for each x in X there is a unique element y in Y with $(x, y) \in f$.*"

Page 30. "*The symbol $f : X \to Y$ is sometimes used as an abbreviation for "f is a function from X to Y."*"

Page 31. "*If the range of f is equal to Y, we say that f maps X onto Y.*" Note: *onto* is used properly, as a preposition.

Page 31. Defines $f(A)$ ("*The notation is bad but not catastrophic*") in the case $A \subseteq$ dom f.

Page 38. Defines $f^{-1}(B)$ in the case $B \subseteq$ ran f and, for a 1-1 function $f : X \to Y$, the usual inverse, also written f^{-1}, "*as the function whose domain is the range of f and whose value for each y in the range of f is the unique x in X for which $f(x) = y$.*" Note: Y is not used in the definition, which thereby is fully general for 1-1 functions.

Page 40. Composition $g \circ f$ defined for f from X to Y and g from Y to Z (restrictive).

[173] Hammack, *Book of Proof*

- *Relation*: page 192. "**Definition 11.7** *A* **relation** *from a set A to a set B is a subset* $R \subseteq A \times B$."

  Page 209, Def. 12.7. Inverse R^{-1} of R from A to B is $\{(y, x) \mid (x, y) : R\}$ (WOLOG).

- *Function*: page 195. "**Definition 11.7** *Suppose A and B are sets. A* **function** *f from A to B (denoted as $f : A \to B$) is a relation $f \subseteq A \times B$ satisfying ....*"

  Page 196, Def. 12.2. B is called the *codomain* of f (conflict).

  Page 198. Some typical reasoning, tacit in most texts, is made explicit. "*Suppose, for example, that $A = \{1, 2, 3\}$ and $B = \{a, b\}$. The two functions $f = \{(1, a), (2, a), (3, b)\}$ and $g = \{(3, b), (2, a), (1, a)\}$ from A to B are equal because the two sets f and g are equal. Observe that the equality $f = g$ means $f(x) = g(x)$ for every $x \in A$. We repackage this idea in the following definition.*
  Definition 12.3 *Two functions $f : A \to B$ and $g : C \to D$ are* **equal** *if $A = C$, $B = D$ and $f(x) = g(x)$ for every $x \in A$.*"
  The author notes that $\{(1, a), (2, a), (3, b)\} = \{(3, b), (2, a), (1, a)\}$ implies $f = g$. The sets A, B, C, D are irrelevant to this conclusion, so "repackaging" should not drag them in. Still, when interpreted correctly (literally), the statement in Definition 12.3 is right — as a theorem! However, due to one of the atavisms in mathspeak, when mathematicians say "if" in a definition, they *really* mean "iff" (if and only if). Why not simply say so? Anyway, if the atavism is assumed, Definition 12.3 entails a logical contradiction with Definition 11.7.

  Page 199, Def. 12.4. *onto* is used as an adjective (assumes B known).

  Page 206, Def. 12.5. Composition $g \circ f$ defined only for $f : A \to B$ and $g : B \to C$ (restrictive).

  Page 210, Def. 12.8. Inverse of 1-1 $f : A \to B$ defined only if $B = \mathcal{R} f$ (restrictive).

[175] Herstein, *Topics in Algebra* (Warning: Herstein uses postfix notation for functions)

- *Relation*: page 11 according to the index, unfortunately not found.
- *Function*: page 10. "*A* **mapping** *from S to T is a subset, M, of $S \times T$ such that ....*"

  Page 10. $\sigma : S \to T$ indicating that σ is a mapping from S to T.

Page 12. *Onto* defined properly, as a preposition, writing *"onto T"*.

Page 12. Equality (re)defined, only for $\sigma : S \to T$ and $\tau : S \to T$ (restrictive).

Page 13. Composition $\sigma \circ \tau$ defined only for $\sigma : S \to T$ and $\tau : T \to U$ (restrictive).

Page 14. Glitch: "onto" used as an adjective in LEMMA 1.2.

Page 14. Inverse defined only for *"a one-to-one mapping of S* onto *T"*.

[191] ISO/IEC, *Quantities and units — Part 2*

- *Function*: page 15. $f : A \to B$ indicating dom $f = A$ and ran $f \subseteq B$.

[194] Jech, *Set Theory*

- *Relation*: page 10. *"An n-ary relation R is a set of n-tuples."* For $n = 2$, Jech uses the term *binary*. Better is *n-adic* and *dyadic relation*, reserving *binary* for a set with two elements (e.g., *binary algebra*).
- *Function*: page 11. *"A binary relation f is a function if $(x, y) \in f$ and $(x, z) \in f$ implies $y = z$."*

  Page 11. *"f is a function from X to Y, $f : X \to Y$, if* $\text{dom}(f) = X$ *and* $\text{ran}(f) \subseteq Y$.*"*

  Page 11. *"If $Y = \text{ran}(f)$, then f is a function* onto *Y."*

  Page 11. Composition $f \circ g$ is considered only for $\text{ran}(g) \subseteq \text{dom}(f)$.

  Page 11. Inverse for a 1-1 function f is defined loosely by $f^{-1}(x) = y$ iff $x = f(y)$.

  Page 11 (bottom). Definitions are generalized to classes.

  Page 12. Term *family* used for a set instead of a function.

[211] Kolmogorov & Fomin, *Introductory Real Analysis*

- *Function*: page 5. Proper use of prepositions, writing *onto N* and *into N*.

  Page 5. Inverse defined only for a one-to-one mapping of *M onto N*.

[213] Krantz, *Real Analysis and Foundations*

- *Relation*: page 19. *"**Definition 1.1** ... A relation* on *A and B is a collection of ordered pairs (a, b) such that $a \in A$ and $b \in B$."*
- *Function*: page 20. *"**Definition 1.2** ... A function* from *A to B is a relation on A and B such that .... We call A the* domain *and we call B the* range.*"* Note: here *range* is what others call *codomain*.

  Page 21, Def. 1.3. *Image*: what others call the *range*.

Page 22, Def. 1.4. A function is *onto* iff its image equals its range.

Page 22,24. Inverse defined only for a function with domain A and "range" B.

[219] Larson & Edwards, *Calculus 9e*

- *Relation*: page 19. "A **relation** between two sets X and Y is a set of ordered pairs, each of the form (x,y), where x is a member of X and y is a member of Y."
- *Function*: page 19. "A **function** from X to Y is a relation between X and Y that ...."

Page 19. Domain and range are defined as usual. No codomain.

Page 21. "A function from X to Y is **onto** if its range consists of all of Y." (conflict!)

Page 25. Defines composition for arbitrary f, g.

Page 343. "A function g is the **inverse function** of the function f if $f(g(x)) = x$ for each x in the domain of g and $g(f(x)) = x$ for each x in the domain of f."

[242] Mendelson, *Introduction to Mathematical Logic*

- *Relation*: page 5. A *relation* is defined as a set of tuples, and a binary relation as a set of pairs, all taken from a single set. Generalization (necessary for functions) is done tacitly. An *inverse relation* is defined as usual.
- *Function*: page 6. "A function f is a binary relation such that $\langle x,y \rangle \in f$ and $\langle x,z \rangle \in f$ imply $y = z$."

Page 6. Domain and range are defined as usual. No codomain.

Page 6. Proper use of prepositions, writing "onto Y" and "into Z".

Page 7. Defines composition for arbitrary f, g.

Page 7. Defines 1-1 by $f(x) = f(y) \Rightarrow x = y$, observes the equivalence with the inverse relation being a function, and defines the latter as the inverse function.

[282] Quine, *Set Theory and Its Logic*

We avoid literally quoting formulae from this text because of its awkward notation, for instance in defining the composition of relations by

'$S \mid R$' for '$\{yx : (\exists z)(ySz \, . \, zRx)\}$'

whereas with the more direct[108] common conventions one could write

$S \mid R = \{(y, x) \mid (\exists z)(y\,S\,z \wedge z\,R\,x)\}.$

Instead, we mention two technical points and a methodological issue.

First, Quine uses the term *dyadic relation*. The term *binary* is better reserved for designating a set with two elements (as in *binary algebra*).

The second point concerns Quine's sound reasons for the order of the arguments when writing "y is the father of x" as $y\,F\,x$ and "y is the square of x" as $y\,S\,x$. The linguistic reason is evident; even "y is smaller than x" is written "$y < x$". More technically, if f is a function, the equality $y = f(x)$ matches $y\,f\,x$. His third reason is making relational and functional coincide notationally: if the relations f and g are functions, $g \circ f = g\,|\,f$ (in fact, one symbol suffices). The reverse convention requires defining $f \circ g = g/f$ as in [312] or $F \circ G = G|F$ as in [316].

Unfortunately, the well-designed convention is far from universal, which raises a methodological issue. In an uncharacteristic diatribe of nearly two pages, Quine denounces the "sorry business" and "glaring perversity" of wantonly changing conventions for the worse, and concludes:

"I have given much space to a logically trivial point of convention because in practice it is so vexatious. The mathematician who switched a seemingly minor point of usage out of willfulness or carelessness cannot have suspected what a burden he created."

This reflects a fitting concern for future generations, which is relevant *a fortiori* for the more serious flaws noted in this bibliography.

[290] Roberts, *Introduction to Mathematical Proofs — A Transition*

- *Relation*: page 176. "A **relation from** A and B is any subset of $A \times B$." Domain and range are defined as usual.

  Page 178. Defines the inverse relation essentially as usual.

  Page 179. Redefines equality of a relation R from A to B and a relation S from C to D, requiring $A = C$ and $B = D$ (conflict).

  Page 180. Composition only for R from A to B and S from B to C.

- *Function*: page 220. "A **function** from A to B, denoted by $f : A \to B$, is a relation from A to B such that ...."

  Page 220. Redefines domain and range (safe). Codomain (conflict).

  Page 223. "**Definition.** *Two functions f and g are **equal**, written $f = g$, if and only if* $\mathrm{Dom}(f) = \mathrm{Dom}(g)$ *and for all* $x \in \mathrm{Dom}(f)$, $f(x) = g(x)$." Note the agreement with the definition on p. 220 and Theorem 1, but the conflict with relational equality p. 179.

  Page 231. "*Onto*" is used as an adjective, (conflict with p. 220).

  Page 240. Defines the inverse of a function $f : A \to B$ as the inverse relation, provided it is a function from B to A. This is restrictive, as it forces f to be onto B.

[295] Royden, *Real Analysis*

- *Function*: page 8. *"By a **function** from X to Y we mean a rule which assigns to each x in X a unique element $f(x)$ in Y."*
  Page 8. Here one finds a brief discussion of the relation to, and possible identification with, the function's graph.
  Page 8. Domain and range are defined as usual. No codomain.
  Page 8. $f : X \to Y$ is used to express that f is a function on X into Y.
  Page 8. Proper use of prepositions, writing "onto Y" and "into Y".
  Page 9. Composition considered only for $f : X \to Y$ and $g : Y \to Z$.
  Page 9. Defines inverse only for a 1-1 function from X onto Y.

[296] Rudin, *Principles of Mathematical Analysis*

- *Function*: page 21. *"Let A and B be sets, and suppose that with each element x of A there is associated an element of B, which we denote by $f(x)$. Then f is said to be a **function** from A to B."*
  Page 21. Domain and range are defined as usual. No codomain.
  Page 21. Proper use of prepositions, writing "onto B" and "into B".
  Page 21. If f is the mapping of A into B and $E \subseteq B$, then $f^{-1}(E) \triangleq \{a : A \mid f(x) \in E\}$. No comment on the domain of f^{-1}.

[300] Scheinerman, *Mathematics — A Discrete Introduction (3rd. ed.)*

- *Relation*: page 73. *"**Definition 14.1** A relation is a set of ordered pairs."*
  Page 74. *"**Definition 14.2** We say R is a relation from A to B provided $R \subseteq A \times B$."*
  Page 74. Inverse of a relation defined as reversing the order in each pair.
- *Function*: page 167. *"**Definition 24.1** A relation f is called a* function *provided $(a, b) \in f$ and $(a, c) \in f$ imply $b = c$."*
  Page 169. **Definition 24.5** defines *domain* and *image* as the set of first and second members of the pairs respectively.
  Page 169. *"**Definition 24.8** Let f be a function and let A and B be sets. We say that f is a* function *from A to B provided $\operatorname{dom} f = A$ and $\operatorname{im} f \subseteq B$. In this case, we write $f : A \to B$."*
  Page 171. **Definition 24.13** defines 1-1 via a set of pairs (as usual) and equivalently by $x \neq y \Rightarrow f(x) \neq f(y)$. **Proposition 24.14**: f^{-1} is a function iff f is 1-1.

Page 172. "**Definition 24.18** *Let* $f : A \to B$. *We say that* f *is* onto B *provided* im $f = B$."
Note: correct, but a missed opportunity: just writing "Let f be a function" is more general. Indeed $f : A \to C$ with im $f = B$ is *onto* B. The "Mathspeak!" note about using "onto" as an adjective is far too indulgent, and may contribute to perpetuating common errors.

Page 183, Def. 26.1. Composition is considered only for $f : A \to B$ and $g : B \to C$. A missed opportunity.

Page 185. "**Proof template 22** *Let* f *and* g *be functions. To prove that* $f = g$, *prove that* dom f = dom g *and prove that, for every* x *in their common domain*, $f(x) = g(x)$." Note: this is a correct characterization of $f = g$, derivable from Def. 24.1.

[302] Shuard, "Does it matter?"

- *Function*: page 8. First version: "*If* A *and* B *are sets, the function* $f : A \to B$ *is a subset of* $A \times B$ *such that* . . . ."
  Shuard considers Cartesian products rather tough material for freshmen, and therefore introduces a "weaker version":
  "*A function* f *consists of two sets* A *and* B *and a rule which* . . . ."
  However, this is not a weaker version, but a definition for a *different kind of object* that has B as an extra attribute.
  Shuard criticizes simpler definitions, in particular Flett [140], for not attaching a codomain, arguing that a codomain is necessary for defining "onto" (as an adjective). Apparently, she overlooks the use of "onto" as a preposition, although Flett [140] sets the example.
  A significant observation is: "*The only analysis book I know, that by Sprecher [306], which uses the domain-codomain definition gets into difficulties later on*". These difficulties concern inverses, an important concept!
  Finally, Shuard mentions that "*Flett's definition wins hands down as far as analysis is concerned.*", yet immediately adds "*In algebra however, it is usually more convenient to start by mentioning the codomain of a function, since we are usually interested in functions which, for instance, map one group into, or onto, another. An isomorphism is often defined as a bijective homomorphism.*"
  This addition is an extremely misleading *non sequitur*. Indeed, the standard convention [191] whereby "*f maps A into B*" means that "*the function f has domain A and range included in B*" is amply adequate for all sensible purposes. For instance, in Herstein [175]:
  [p. 46]"DEFINITION. *A mapping* ϕ *from a group* G *into a group* $\overline{G}$ *is said to be a* homomorphism *if for all* $a, b \in G$, $\phi(ab) = \phi(a)\phi(b)$."
  [p. 49]"DEFINITION. *A homomorphism* ϕ *from a group* G *into* $\overline{G}$ *is said*

to be an isomorphism *if ϕ is one-to-one."* (not always onto $\overline{G}$). If one wants ϕ to be a bijection onto $\overline{G}$, saying so suffices.

[304] Smith, Eggen & St. Andre, *A Transition to Advanced Mathematics*

- *Relation*: page 135 *"**Definition** Let A and B be sets. R is a **relation from** A **to** B iff R is a subset of $A \times B$."*
  Page 137. *"The **domain** of the relation R from A to B is the set*
  $\quad \text{Dom}(R) = \{x \in A: \text{there exists } y \in B \text{ such that } x\,R\,y\}"$
  (similarly for the range). Note: clearly "from A to B" is restrictive. Much better is the subsequent clarification, *"Thus the domain of R is the set of all first coordinates of ordered pairs in R."*
  Page 139. Defines the inverse of a relation R as usual (order reversal).
  Page 141. Composition is considered only for R from A to B and S from B to C.

- *Function*: page 186 *"**Definitions** A **function** (or **mapping**) **from** A **to** B is a relation f from A to B such that . . . .*
  *We write $f : A \to B$, and this is read "f is a function from A to B". The set B is called the **codomain of** f."*
  Page 189. *"Because functions are sets of ordered pairs, we may say that functions f and g are **equal** if and only if $f \subseteq g$ and $g \subseteq f$."* Referring to an example: *"Although f and g may have different rules and different codomains, they have the same domain and are both equal to* [the same set of pairs]. *Therefore $f = g$."* This leads to
  *"**Theorem 4.1.1** Two functions f and g are equal iff*
  (i) $\text{Dom}(f) = \text{Dom}(g)$ and (ii) for all $x \in \text{Dom}(f)$, $f(x) = g(x)$."
  The observation that functions may have different codomains without preventing equality amounts to recognizing that the codomain is not an attribute of the functions but, at most, of the context.
  Page 196. Defines the inverse of a function as the inverse relation, and observes that it might not be a function.
  Page 205. *A function $f : A \to B$ is **onto** B iff $\text{Rng}(f) = B$.*
  Note: correct, but a missed opportunity; writing "A function f is **onto** B iff $\text{Rng}(f) = B$" is more general.
  Page 215. *"**Theorem 4.4.2** Let F be a function from set A to set B.* **(a)** *F^{-1} is a function from $\text{Rng}(F)$ to A iff F is one-to-one.* **(b)** *If F^{-1} is a function, then F^{-1} is one-to-one."* Observe: no loss of generality.

[307] Stewart, *Calculus: Early Transcendentals (7th. ed.)*

- *Function*: page 10. *"A **function** f is a rule that assigns to each element in a set D exactly one element, called $f(x)$, in a set E. The set D is called* the **domain** of the function." Furthermore, "The **range** of f is the set of all possible values of $f(x)$ as x varies throughout the domain." Note: no "codomain"!

Page 11. The **graph** of a function is defined as usual.

Page 40. "The **composite function** $f \circ g$ (also called the **composition** of f and g) is defined by $(f \circ g)(x) = f(g(x))$. The domain of $f \circ g$ is the set of all x in the domain of g such that $g(x)$ is in the domain of f."

Page 385. "Let f be a one-to-one function with domain A and range B. Then the **inverse function** f^{-1} has domain B and range A and is defined by

$$f^{-1}(y) = x \Longleftrightarrow f(x) = y \text{ for any } y \text{ in } B."$$

[312] Suppes, *Axiomatic Set Theory*

- *Relation*: page 57.
  "DEFINITION 1. A is a relation $\leftrightarrow (\forall x)(x \in A \rightarrow (\exists y)(\exists z)(x = \langle y, z\rangle))$." In other words, a relation is a set of (ordered) pairs.

  Page 58. Domain $\mathcal{D} R$ and range $\mathcal{R} R$ of a relation R as usual.

  Page 63. Composition of relations is called *relative product*:
  "DEFINITION 7. $A/B = \{\langle x, y\rangle : (\exists z)(x \, A \, z \, \& \, z \, B \, y)\}$."

- *Function*: page 86.
  "DEFINITION 39. f is a function $\leftrightarrow f$ is a relation $\& (\forall x)(\forall y)(\forall z)(x \, f \, y \, \& \, x \, f \, z \rightarrow y = z)$."
  In other words, a function is a functional relation.

  Page 87. Composition of functions: "DEFINITION 41. $f \circ g = g/f$."

  Page 88. "DEFINITION 42. f is 1-1 $\longleftrightarrow f$ and $\breve{f}$ are functions."

  Page 88. "DEFINITION 43. f is 1-1 $\longrightarrow f^{-1} = \breve{f}$."

[316] Tarski & Givant, *A Formalization of Set Theory Without Variables*

- *Relation*: page 3. "By a (binary) relation *we understand a class of ordered pairs*."

  Page 3. Domain DoR and range RnR of a relation R as usual.

  Page 3. Composition, called *relative product*:
  $R|S = \{\langle x, y\rangle : \text{ for some } z, x Rz \text{ and } zSy\}$.

- *Function*: page 3. "A function *is a relation F such that* ....."

  Composition: $F \circ G = G|F$, together with the observation that $(F \circ G)x = F(Gx) = FGx$ for every $x \in Do(F \circ G)$.

  Note: writing FGx for $F(Gx)$ is unwise because, in view of higher-order functions, FGx is better reserved as shorthand for $(FG)x$.

Page 3. *"If both F and F^{-1} are functions we say that F is* one-to-one *and refer to F^{-1} as the* inverse *of F."*

[330] Velleman, *How To Prove It: A Structured Approach*

- *Relation*: page 171. **"Definition 4.2.1** *Suppose A and B are sets. Then a set $R \subseteq A \times B$ is called a* relation *from A to B."*

  Page 172. $\mathrm{Dom}(R)$ and $\mathrm{Ran}(R)$ defined as usual. No codomain.

  Page 173. **"Definition 4.2.3** [...] *Finally, suppose R is a relation from A to B and S is a relation from B to C. Then the* composition *of S and R is the relation $S \circ R$ from A to C defined as follows:*
  $$S \circ R = \{(a,c) \in A \times C \mid \exists b \in B((a,b) \in R \text{ and } (b,c) \in S)\}.$$
  Notice that we have assumed that the second coordinates of pairs in R and the first coordinates of pairs in S both come from the same set B. If these sets were not the same, the composition $S \circ R$ would be undefined."

  The latter stipulation is highly misleading in suggesting that $S \circ R$ can be undefined. Indeed, even if $R \subseteq A \times B$ and $S \subseteq B' \times C$ with $B \neq B'$ we also have $R \subseteq A \times (B \cup B')$ and $S \subseteq (B \cup B') \times C$, so $S \circ R$ is captured by Definition 4.2.3 without any issues.

- *Function*: page 226. **"Definition 5.1.1** *Suppose F is a relation from A to B. Then F is called a* function *from A to B if for every $a \in A$ there is exactly one $b \in B$ such that $(a,b) \in F$. To indicate that F is a function from A to B, we will write $F : A \to B$."*

  Page 229. **"Theorem 5.1.4.** *Suppose f and g are functions from A to B. If $\forall a \in A(f(a) = g(a))$, then $f = g$."* Mind the gaps!

  Page 230. *"If $f : A \to B$ and $g : B \to C$, then f is a relation from A to B and g is a relation from B to C, so it makes sense to talk about $g \circ f$, which will be a relation from A to C. In fact, it turns out to be a function from A to C as* [**Theorem 5.1.5**, page 231] *shows."* Note: this is restrictive because the above reasoning for $S \circ R$ leads to the requirement $\mathrm{Ran}(f) \subseteq \mathrm{Dom}(g)$ as explained in Subsection 7.2.5.

  Page 236. **"Definition 5.2.1.** *Suppose $f : A \to B$. We say that f is* onto *if $\forall b \in B \exists a \in A(f(a) = b)$."* As expected, this conflicts with Definition 5.1.1.

  Page 245. **5.3 Inverses of functions** This section considers only the case where the inverse of a function from A to B is a function from B to A. This is restrictive by requiring the function to be not just 1-1 but also onto in the sense of Definition 5.2.1.

[334] Walter D. Wallis, *A Beginner's Guide to Discrete Mathematics*

- *Relation*: page 93–94. Informal introduction, concluding with: "*Formally, a binary relation from S to T can be defined as a subset of S × T.*" Page 94. Composition is considered only for τ from R to S and σ from S to T, and denoted by $\tau\sigma$.

- *Function*: page 106. "*A function, or mapping, from a set S to a set T is a special kind of relation from S to T, one in which each element of S occurs as the first element of* precisely *one ordered pair. The set S is called the* domain *of the function and the set T its* codomain." [...] For an arbitrary relation, "*we would write f ⊆ S × T, but more often when dealing with functions we write f : S → T*".

  Page 106. Composition, written $g(f)$, is defined for any f, g, with $\mathcal{D}(g(f))$ as expected. Note: $g(f)$ is poor notation for composition, and is even disastrous for higher-order functions.

  Page 107. "*We say that f is* onto *if f(S) = T.*" It is tacitly assumed that f is declared as $f : S \to T$.

  Page 108. **Inverse Functions** After defining the inverse of a function as the inverse relation, this section proceeds to "***Theorem 26.** Let f be a function f : S → T. Then f is one-to-one and onto if and only if its inverse relation f^{-1} is a function from T to S.*" The more general case where f is just 1-1 is not mentioned.

[348] Elias Zakon, *Mathematical Analysis, Vol. I*

- *Relation*: page 8. "*It is customary to call any set of ordered pairs a* relation. *Instead of $(x, y) \in R$, we also write xRy.*"

  Page 8. "*Since relations are sets, equality R = S for relations means that they consist of the same elements (ordered pairs).*"

  Page 9, Def. 1. Domain and range, written D_R and D'_R.

  Page 9, Def. 2. The set $R[x]$ (R-relatives of x): $y \in R[x]$ iff xRy. Note: an argument for the convention of Peano, Gödel and Quine.

- *Function*: page 10. "***Definition 3.** A relation R is called a* mapping, *or a* function *iff every element $x \in D_R$ has a* unique *R-relative, so that $R[x]$ consists of a* single *element [...] denoted by $R(x)$*"

  Page 10. The inverse of a map (or function) is defined as the inverse relation, and a footnote mentions that it is a map only if the given function is 1-1.

  Page 11. "*A mapping f is said to be "from A to B" iff $D_f = A$ and $D'_f \subseteq B$; we then write f : A → B. If, in particular, $D_f = A$ and $D'_f = B$, we call f a map of A* onto *B.*"

  Note: since functions are relations, equality obeys the same criterion.

7.4   Some ramifications

The concept of a *function* as characterized by Apostol's one-liner [Definition 7] is unlikely ever to be surpassed in simplicity and in its universality throughout mathematics, and therefore can safely be considered the definitive design. Of course, this observation must be interpreted modulo the abstraction step leading to Definition 12, which is essentially just a change of perspective covering the same concept. The universality is apparent from the many properties listed in Subsection 7.2.7 and additional ones in [43, 244].

By contrast, the notion of *codomain* is spurious: it merely *appears* in some definitions (sound and unsound), but nowhere in mathematics at large or in applications does it seem to be actually *used* in any constructive or interesting way. Since it also destroys the aforementioned properties and hence the universality of the function concept, it might at best have some benefits, yet to be discovered, in some niche area. Hence, when introducing codomains in any area, proper justification is most imperative.

Apart from this, codomains seem to be sufficiently confusing for many authors as to cause unsoundness in textbooks; one can imagine the impact on students.

Half a century ago, the function concept was free of issues, and present-day calculus texts also preserve the aforementioned simplicity and universality. Apparently, over the years, too many cooks have thoroughly spoiled the broth.

Educational considerations   When I described the current state of affairs during a presentation at the *MAA Mathfest 2015* [44], some people in the audience asked whether students experience any problems because of the defective definitions.

My initial answer was that students who accept defective definitions at face value do not understand these definitions at all, and hence have a huge problem. Afterwards it became clear that the people asking this question were thinking about difficulties with homework and exams.

My answer should therefore have been that instructors using the defective definitions must themselves be unaware of the issues and therefore are unlikely to give homework or exam questions that might expose any deficiencies in the textbook. Indeed, it appears from the earlier annotations to [120] that even the most diligent authors unconsciously or consciously skirt these issues, although the treatment of functions in *all* texts using codomains becomes very noticeably chafing at one point or another. No subtleties are involved at any point.

Apparently the very existence of the term *codomain* makes some instructors feel that they must use it. They need not, but should at least account for it, since students are bound to meet the term in other texts.

One possibility consists in defining *codomain* as (yet another) synonym for *range*, which would yield an elegant symmetry with *domain*. However, very few authors (references lost) have chosen this option and the damage has gone so far that this choice would only add to the confusion.

Hence the most sensible way to resolve the problem is to refrain from complicating the function concept beyond Definitions 7 or 12, and mention the *codomain* only in a series of critical questions and exercises in the style of Exner [120], and to expose deficiencies in definitions rather than skirting them. The detailed quotations in Section 7.3 are a very good source of inspiration for such exercises.

In addition, it is most helpful to elucidate the worded statements in definitions not just by examples but also by judicious use of symbolism to the degree illustrated in the earlier quotes from Suppes [312], although in a more carefully designed and hence more palatable notation. Arguably, if textbooks of the type *Introduction to Proof* were a little less reticent about symbolism, unsound definitions would have been nipped in the bud.

Relations and functions in category theory In category theory, the *disjointness axiom* for hom-sets as stated, for instance, by MacLane [232, p. 27] calls for codomains and hence leads to the shortcomings outlined in Subsection 7.2.7 and in particular, a function concept that is unacceptably restrictive for practical areas of mathematics such as calculus. Even composition, the basic concept in category theory, is not defined for arbitrary pairs of relations or functions.

Hence, the disjointness axiom is also an impediment from a theoretical viewpoint: an axiomatization that is unable to cover the simplest and most universal concept of a *function* can hardly be considered exemplary in any positive sense (elegance, unifying power etc.). Yet, although MacLane [232, p. 27, last sentence] suggests that the disjointness axiom is not essential to the basic ideas of category theory, all available texts uncritically adopt it, impairing the relevance to mathematics at large.

Fortunately, an initial study reveals that this axiom is only a matter of tradition, probably originating from a particular theoretical niche area, and can be omitted with very minor tweaks. Further elaboration of this topic goes beyond the scope of this bibliography and is presented in a separate paper.

Relations and functions in computer science   Many texts on the *theory of programming*, including [244], use relations and functions in the standard sense [Definitions 1 and 7]; others attach codomains, rarely in an unsound way, but clearly leading to the shortcomings outlined in Subsection 7.2.7.

In programming *languages*, the term *function* has long been abused by allowing side-effects, so here we consider only *pure functions* which have no side-effects and hence can be modeled as mathematical functions. Often the model must assume codomains because compilers handle function types in a manner based on simple algorithms developed decades ago. Automatic handling of function types in the standard sense [Definition 8] requires the capabilities of a *verifying compiler*, currently still one of the Grand Challenges in computer science.

For the same reason, most so-called *functional* programming languages assume the variant with codomains, in view of the type inference algorithms available.   It must be said here that most functional programming languages also miss other important opportunities.

Indeed, one would expect that a functional programming language unifies different kinds of mathematical objects by capturing them as functions to the greatest extent that is conceptually advantageous, but this has not (yet) happened.   For instance, sequences and lists are conveniently viewed as functions in mathematics [172, 213, 295, 296], while in functional programming, lists are defined as a recursive data type [34, p. 7].

This fallback from the functional view is regrettable conceptually, but also has practical drawbacks: it requires the introduction of a plethora of list-processing functions for purposes that are already captured by the more universal basic operators mentioned in Subsection 7.2.7 and in [43, 244].   For instance, if $\{x, y\} \subseteq \mathcal{D}f$, then $f \circ (x, y) = (f\,x, f\,y)$, which obviates map and its two variants *listl* and *listr* [34, p. 9]. Similarly, if $x \in \mathcal{D}f \cap \mathcal{D}g$, then $(f, g)^\mathsf{T} x = (f\,x, g\,x)$ which obviates and generalizes pair [34, p. 266].   Note the duality with composition.   One last example is $((x, y), (a, b), (3, 4))^\mathsf{T} = ((x, a, 3), (y, b, 4))$, which obviates and generalizes zip [34, p. 268].

Similar considerations hold for specification languages. The relation between specification languages and theorem provers or proof assistants is somewhat similar to the relation between programming languages and compilers. An interesting discussion is found in [217].

In the interests of universality and of acceptability in mathematics at large and in engineering, it is important that the design of specification languages should aim to be close to the language of mathematics at large rather than accepting restrictions based on implementation considerations for the tools (theorem provers, proof assistants). The tools should support the concepts, not vice versa.

For instance, Lamport's TLA$^+$ fully preserves the simplicity and universality of the standard function concept by formulating Definition 12 and Definition 8 (the ISO standard) as follows [216, p. 48].

> A function has a domain, written DOMAIN f, and it assigns to each element x of its domain the value $f[x]$. (Mathematicians write this as $f(x)$, but TLA$^+$ uses the array notation of programming languages, with square brackets.) Two functions f and g are equal iff they have the same domain and $f[x] = g[x]$ for all x in their domain.
> The *range* of a function f is the set of all values of the form $f[x]$ with x in DOMAIN f. For any sets S and T, the set of all functions whose domain equals S and whose range is any subset of T is written $[S \rightarrow T]$.

Our only criticism concerns the wasteful use of square brackets — a scarce resource!

Lamport observes that "No model checker can handle all the specifications that we can write in a language as expressive as TLA$^+$. However, TLC seems able to handle most TLA$^+$ specifications that people actually write." The capabilities of the automated tools for TLA$^+$ are regularly expanded, and updates are available via the website *http://research.microsoft.com/en-us/um/people/lamport/tla/tla.html*.

Arguably, it is wise to avoid model checkers, theorem provers or proof assistants that sacrifice the universality of the standard function concept. Indeed, once cast in the concrete of software support, poor design decisions in definitions are difficult to reverse. Especially in education, software tools represent a great danger of undermining the critical spirit of students, causing misconceptions and creating bad habits; choices do matter.

7.5   Conclusions

In view of the many references giving proper accounts for *relations* and *functions*, it is regrettable that so many unsound treatments have emerged in recent years. If Quine already has *"given much space to a logically trivial point of convention because in practice it is so vexatious"* [282, p. 25], conceptual flaws require an even more extensive analysis of the literature.

For starters, this analysis invalidates the myth that the existence of different definitions just reflects the different needs in various areas of mathematics. Our sample draws from logic, set theory, analysis/calculus, discrete mathematics, algebra and proof methods, and yet all definitions are logically equivalent by Theorem 3 (except the patched variant in [36]). The real differences reflect the care devoted to conceptualization and formulation. Ironically, most analysis/calculus texts define the *function* concept properly and with full generalizability in the first few pages, whereas in most *introduction to proof* texts one has to wait for over 100 pages, only to be disappointed by flawed definitions.

Another observation is that confusion evidently arises from the convoluted formulation of Definition 13. To paraphrase Quine [282, p. 25]: *"The mathematician who switched a seemingly minor point of formulation out of willfulness or carelessness cannot have suspected what a burden he created."* Indeed, all typical misconceptions can be explained as ramifications of Definition 13 that are either illusory or overlooked. An illusion is that Definition 13 makes Y a function attribute. An oversight is not understanding that the standard interpretation [191] of $f : X \to Y$ amply suffices for all sensible purposes *without* making Y a function attribute. Attaching a codomain to a function yields a different function concept [84, 140], without serving any useful purpose and to the detriment of universality. This is unacceptably restrictive in practice. Evidently the obligation to justify definitional design decisions applies doubly here.

The upside of the preceding observations is that the way to eliminate all problems, educationally and practically, is simple: by avoiding Definition 13 and starting from the simple yet logically equivalent Definition 7 or the conceptually equivalent Definition 12.

I humbly confess having vastly underestimated the time required for putting together Section 7.3. Yet, precisely for this reason, it may save students, instructors and authors a multiple of that time collectively.

8. Conclusions

In 2006 I (Edgar Daylight) defended my Ph.D. thesis at KU Leuven. Dr. Jeroen Voeten from the Netherlands was in my defense committee. He had shared with me a few days earlier some of the "fundamental limitations" he had found for transformation systems, such as the system I had designed, implemented, and documented for my Ph.D. defense. Voeten had already described his theoretical insights in a published article, entitled: 'On the fundamental limitations of transformational design' [333], and he wanted me to incorporate his findings into my Ph.D. dissertation.[109]

I had to follow a Master's program in Logic at the University of Amsterdam (2007–2009) to actually start understanding each and every argument in Voeten's theoretical paper. My incentive to do so was because I intuitively felt that Voeten's reasoning was flawed. In the process, and due to my strong interest in the history of ideas, I started to see that several computer scientists, including the most prominent ones, had been making and were continuing to make category mistakes in the same way as Voeten.

Very briefly, Voeten's point was that he had mathematically *proved* that certain *systems* cannot be *engineered*. In sociological terms, mathematics, and computability theory in particular, prevails over engineering; all engineers had better start listening to mathematicians. I believe I am today in a position to improve Voeten's wording with regard to his own article. I would say that:

> If I model my transformation system with his specific Turing-machine language, then I obtain a classical undecidability result from computability theory which potentially gives me some insights into the industrial problem that I and fellow engineers are trying to solve.

The undecidability result does not tell me that specific systems cannot be built. Perhaps an engineer could take the undecidability result into account in some way, when engineering a real system. Appendix A contains my detailed scrutiny of Voeten's article.

While studying the history of ideas in computer science, I also started to notice that philosophers such as James Moore, James Fetzer, and Timothy Colburn, along with the sociologist Donald MacKenzie, had made similar observations years and even decades before me. However, computer scientists have mostly ignored these writings, or, as in the case of Fetzer, have ridiculed him. I am of course referring to Fetzer's 1988 article in the Communications of the ACM, entitled: 'Program Verification: The Very Idea' [125]. The story is told in MacKenzie's 2004 book *Mechanizing Proof* [233] and in my opinion should be mandatory reading material for students and academics alike.

I, too, was a hard-core formal methodist during my Ph.D. years. Interestingly, none of my peers ever told me that the papers I was reading (i.e., the writings of Tony Hoare and others) were flawed in some crucial respects — as Fetzer already remarked in 1988. My educated guess is that even top computer scientists are not well aware of the history of their own discipline.

Perhaps many scientists in the Western world believe that scientific progress amounts to being able to give one precise answer to each fundamental question, such as the following:

What is a computer program?

But wait a minute. Dijkstra says in 1973 that a program is a mathematical object of finite capacity, while Christopher Strachey says, in that very same year, that a program is a mathematical object of infinite size. How can this be? Perhaps computer science is not so clear-cut after all. Moreover, Peter Naur insisted that a program is a model of the real world, which is *not* a logico-mathematical construction. Thus, here we already have three conflicting views on computer science's most basic question; as I have also shown, this pluralism is also very much present today. These observations might suggest that computer science is still too young a field. Another possibility worth contemplating is that computer science is intrinsically pluralistic in nature and that it will remain so in the distant future.[110]

8.1   Three Turing Tales

Coming to the title and a summary of this book, *Turing Tales* refers primarily to the following three rectifications.

1. **Tatty Turing Tale:** Turing did not invent the computer, nor did his 1936 paper lead to the invention of the computer. Turing's universal machine concept was re-appropriated in the field of automatic programming in the 1950s. However, even without Turing, automatic programming would have flourished.

2. **Tiny Turing Tale:** While Turing's 1936 paper did not influence computer building at large, it had exceptional influence in a tiny country, The Netherlands, where Willem van der Poel in Delft, and later in The Hague, built simple computers based on firm, logical principles and on Turing's 1936 notion of universality in particular. Striving for Turing universality in a practical setting led van der Poel to develop intrinsically slow machines which he had to program in a very machine-dependent manner in order to make the complete package industrially viable. Van der Poel perfected optimum coding and underwater programming, i.e., two techniques that Turing, too, had embraced as a programmer.

3. **Titanic Turing Tale:** The term 'Turing completeness' and related terms (such as 'Turing machine' and 'Turing universality') are abused everywhere, *including* in clitist quarters of academic computer science. I have provided several examples of published category mistakes that need rectification for the sake of common understanding, i.e., mistakes in connection with the halting problem, computer viruses, formal verification, and so on.

To re-convey the aforementioned tales in some more depth, I start with the dialogue in Figure 8.1, which I have used in my course on logic and computation at Utrecht University. The philosophy students in my class see the category mistake in the dialogue from day one, while almost none of my computer science students do. By the last lesson, half of my computer science students have come to see the mistake, while the other half remain *true* computer scientists.

If I were to have a similar discussion today as in Figure 8.1, I would stress that if my laptop has an unbounded memory capacity, *even then* a distinction has to be made — *at least once in writing* and for the sake of common understanding — between the concrete physical object and the abstract object. I can destroy my laptop along with the computer

Computer Scientist: I can't find a proof that Howard Aiken's machine from 1944 was Turing universal. Can you give me a paper in which this claim is proved? If not, I will have to look up the machine specifications and prove Turing universality myself.

Philosopher: Proving that Aiken's machine was Turing universal is not possible mathematically, just like it is impossible to prove that my laptop is equivalent to a universal Turing machine. I can model my laptop, or any other physical device, as a Turing machine or as a finite state machine or as something else. In each case I will have to abstract away several details from my device. If I choose a finite-state-machine model, then I can argue that my laptop is not as powerful as a universal Turing machine. If I model my laptop as a universal Turing machine, then I can view my laptop as Turing universal.

Computer Scientist: I'm not sure I agree, it should be rather easy to prove Turing universality of a given computing machine M.

Philosopher: One can prove that two models of computation are equivalent or in-equivalent. However, you can't prove that a mathematical model is equivalent to a physical machine M.

Computer Scientist: But all you need to do is program a universal Turing machine in M. In doing so, you may assume that M's memory capacity is unbounded. Proving Turing universality is thus no big deal.

Philosopher: You have to make an assumption about $M's$ memory capacity. You could just as well have taken the opposite assumption in your modeling to "show" that M is *not* Turing universal.

Computer Scientist: If I am not allowed to make *that* assumption, then not a single machine is Turing universal. So the assumption seems necessary. Because the assumption is so simple, it seems to be the right assumption to make.

Philosopher: Assuming that machine M has a bounded memory is also simple.

Figure 8.1: A 2014 discussion between an eminent computer scientist and the present author, i.e., the "philosopher." The computer scientist wants to prove something mathematically about an engineered system. Specifically, he fuses the abstract world of Turing machines and partially computable functions with the world of concrete physical objects. The latter contains Aiken's 1944 machine and the operations that it performed in a real, physical world.

program residing in it. However, I cannot destroy the specific universal Turing machine that both actors in the dialogue are discussing.

The "computer scientist" in Figure 8.1 is a prominent researcher, and I have illustrated with several other examples in this book that his faulty reasoning is rather typical and not at all exceptional. Furthermore, it is easy to find research papers and books that use this faulty kind of reasoning to make claims about Turing's allegedly important role in the history of the modern computer. For instance, a conflation of the universal machine and the stored-program computer leads to the following statement:

> Turing invented the computer.

This is a specific example of a compact, yet incorrect, formulation. Another example is:

> During the 1950s, a universal Turing machine became widely accepted as a conceptual abstraction of a computer.

However, as discussed in this book, the computer builders and programmers who appropriated ideas from Turing's 1936 paper in the 1950s were few in number.

Compact and appealing formulations often lead to wide disseminations of myths. It is much better, I believe, to make specific claims about multiple, well-chosen actors, such as Saul Gorn in the USA and Willem van der Poel in The Netherlands and to compare and contrast their views with regard to the same technological or theoretical object.

On the one hand, both Gorn and Van der Poel did essentially view a universal Turing machine as a conceptual abstraction of a computer in the 1950s, and in contrast to many computer scientists *today*, Gorn and van der Poel used the universal Turing machine *solely* as a mathematical *model* of a general-purpose computer and thus not as a *synonym* for a computer. On the other hand, there is also a large difference between Gorn and van der Poel and this lies in the way they appropriated and applied the concept of the universal Turing machine in their engineering work. Gorn embraced the potential infinite, while van der Poel only wanted to work in a strict finite setting (in which, as he emphasized in his Ph.D. dissertation, everything is trivially decidable).

In Amsterdam, Adriaan van Wijngaarden and Dijkstra were not influenced by Turing's writings [98]. Moreover, in the same way as van der Poel, Dijkstra preferred to work with finite state machines.

However, van Wijngaarden and Dijkstra also abstracted away from physical computer machinery, in the following sense: they programmed in a machine-independent manner (and were quite successful in this regard). Van Wijngaarden and Dijkstra thus had much more in common with Gorn, John Carr, and Alan Perlis in the USA than with van der Poel in their own country.

Computer Programs and Mathematical Programs

Figure 8.1 illustrates only the categorical distinction between computers and mathematical models, while this book is also about the less obvious categorical distinction between computer programs and mathematical programs. Recall that a mathematical program is used as an abbreviation in this book for a mathematical model of a computer program.

A recurring example concerns Dijkstra's and Christopher Strachey's views on programming languages in 1973. By then, Dijkstra was viewing both a computer and a computer program as mathematical objects of *finite* size. By contrast, Strachey in England worked with infinite precision real numbers and appropriated more ideas from modern logic than Dijkstra.

Neither Dijkstra or Strachey always explicitly distinguished — or, perhaps, they rarely ever explicitly distinguished — between the computer program on the one hand and their mathematical program on the other hand. Today this conflation between computer programs and mathematical programs is very much present, as illustrated by the following Wikipedia excerpt:

> In computability theory, Rice's theorem states that all non-trivial, semantic properties of programs are undecidable. A semantic property is one about the program's behavior (for instance, does the program terminate for all inputs), unlike a syntactic property (for instance, does the program contain an if-then-else statement). A property is non-trivial if it is neither true for every program, nor for no program.

> We can also put Rice's theorem in terms of functions: for any non-trivial property of partial functions, no general and effective method can decide whether an algorithm computes a partial function with that property. Here, a property of partial functions is called trivial if it holds for all partial computable functions or for none, and an effective decision method is called general if it decides correctly for every algorithm.

In contrast to the second paragraph, the first paragraph in this Wikipedia excerpt deserves further scrutiny. The comments made in the first paragraph can only be correct if the following two conditions hold:

- a "program" denotes a mathematical program and,
- the mathematical program is described with a Turing-complete language.

Making these assumptions explicit is what a large part of this book is about. More precisely, and in line with with some recent developments in the philosophy of technology, I have made a distinction between the following three categories:

1. computers (including laptops and iPads) are concrete physical objects,

2. computer programs and also computer programming languages are technical artefacts, and

3. mathematical programs, mathematical programming languages, Turing machines, finite state machines, and prime numbers are abstract objects.

Through the consistent use of these categorical distinctions, I have shown that computer scientists often refer to a computer program when they are actually referring to *a* mathematical model of the computer program. In fact, they are referring to the program text of the mathematical program under scrutiny, rather than the computer program itself. Figure 8.2 provides a visualization of these crucial distinctions, which were introduced in Chapter 5 and discussed at length in Chapter 6.

When computer scientists refer to their mathematical program, they often incorrectly believe they are also directly referring to the actual computer program. Specifically, they do not distinguish between different representations of the mathematical program, such as a program text and a computer program. Each representation is a technical artefact and not a mathematical object.

Figure 8.2 thus depicts one abstract object and two objects that are not abstract: the program text and the computer program.[111] In line with recent developments in philosophy, I call these latter two objects technical artefacts. No harm is done, however, if the reader chooses to view both the program text and the computer program as concrete physical objects; doing so will not undermine my main complaint, which can now be generalized as follows:

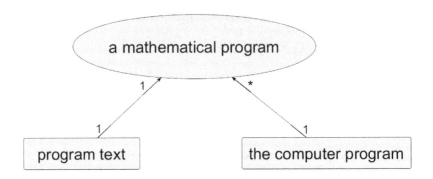

Figure 8.2: The computer program at hand can be modeled using one or more mathematical programs. One such model is depicted, along with its program-text representation. Specifically, the left arrow denotes the relationship 'is the textual representation of' and the right arrow denotes the relationship 'is the electronic representation of'.

> Computer scientists do not consistently distinguish between some of their favorite abstract objects and the representations thereof in the real world.

8.2   John Reynolds and Proof-Carrying Code

A final example of a category mistake comes from a 2012 letter which John Reynolds wrote to me [288]. On the one hand, the letter contains an extensive amount of valuable information about the history of programming languages. On the other hand, I did experience some difficulty comprehending the following part of Reynolds's letter because I had by then already studied MacKenzie's 2004 book *Mechanizing Proof* [233] in some detail. Here are Reynolds's words pertaining to C computer programs:

> As an example, it is likely that, in perhaps ten years time, when you download a **program** from the Internet, you will also download a formal proof that the **program** will respect safety conditions (e.g. no buffer overflow or dereferencing of pointers into the wilderness) that will ensure that it cannot disrupt the behavior of other **programs** running simultaneously. And your computer will check these proofs before running the **program**. [288, my emphasis]

Later on, Reynolds writes that when this notion of "proof-carrying code" is "complete and widely used, logic will have won a major victory." However, again, the claims made in the excerpt can, strictly speaking, never hold. For instance, the first and second occurrences of the word "program" refer to very different objects. The first occurrence refers to a computer program, while the second occurrence refers to *a* mathematical program. At best, it is the mathematical program that will respect *mathematically modeled* safety conditions, not the real program residing in someone's computer.

Spelling out these trivialities might give the false impression that programming language specialists are completely unaware of, or would strongly disagree with, these clarifications. I don't think this is the case. I suspect these experts simply do not spend much time scrutinizing the *informal* aspects of the topic at hand, at least not in published writings.

Coming back to Reynolds's usage of the word "program," a similar remark holds for the third and fourth occurrence of the word in the previous excerpt. Specifically, the "behavior of other programs" refers to mathematically modeled behavior, i.e., to the behavior of other *mathematical* programs. The fourth and final occurrence of "program," by contrast, refers to the *computer* program at hand.

However, this is not all. A formal proof is represented electronically inside a computer. So it is an electronic representation of the formal proof that will be queried, not the formal proof itself, nor a textual representation of it.

Here, then, is my humble attempt to add adjectives and nouns such as "mathematical" and "computer," to Reynolds's excerpt with the sole purpose of trying to fully comprehend Reynolds's view on proof-carrying code with regard to C computer programs:

> As an example, it is likely that, in perhaps ten years time, when you download a *computer* program from the Internet, you will also download an *electronic representation* of a formal proof that the *mathematical* program (which is *a model* of the *computer* program under scrutiny) will respect mathematically modeled safety conditions (e.g. no buffer overflow or dereferencing of pointers into the wilderness).
>
> This will *not* ensure but will definitely increase our *confidence* that the *computer* program cannot disrupt the behavior of other *computer* programs running simultaneously. And your computer will use yet another — albeit much simpler —

> *computer* program to check *electronic representations* of these
> proofs before running the *computer* program at hand.

The above is me thinking out loud after having received critical remarks from a professional safety engineer in industry. (I am currently working as a consultant and formal methodist in the automotive industry.) Just like Reynolds, I, too, have recently been putting mathematics in the driver's seat. In safety engineering, however, one is kindly requested to substitute the following words used by Reynolds in order to be taken seriously in daily discourse:

1. "will respect safety conditions" –> "shall respect safety conditions"

2. "ensure" –> "increase our confidence"

Indeed, books on hazard and risk analysis will confirm that requirements always need to be expressed with "shall" and not "will." The rationale for this seemingly small correction is that we, formal methodists, do not have any guarantee whatsoever that the safety conditions *will* be met at runtime by the engineered *system* even though we have obtained guarantees — or have come very close to obtaining guarantees — about some mathematical *model* of the system. Increasing our confidence in the system's correctness is the best we can do.

In retrospect, then, I find it extremely ironic that non-formal-methods people have to persuade formal-methods advocates, like myself, when working in industry, to be *more precise* in the way we formulate our mathematical findings. Professional safety engineers who are responsible for, say, the software in an airplane are rightly concerned with minute details which apparently go unnoticed by formal methodists.

Another major, and strongly related, technical point that I have tried to convey in this book is that many programming language specialists want to attach precisely one meaning to each computer programming language. This is, however, not the most intuitive thing to do from an engineering perspective. If an engineer is put in the driver's seat, then he or she will model a computer and a program residing in that computer in many complementary ways. The richness lies in the multitude of models. People like Dijkstra and Strachey each attached a different semantic model to the same technical artefact and similar things are happening today in the world of C computer programs. Like it or not, C is informal, it is made (by humans), and we can mathematically model it in many ways.

In this regard, I mention Raymond Boute's writings on System Semantics in the 1980s [46, 47]. His work has come to my attention right before this book went to press. While denotational semantics "defines only a single interpretation in terms of an abstract (mathematical) domain," Boute explains, "system semantics introduces various models pertaining to properties of concrete (physical) systems" [46, p.1219]. In future work I hope to explore this avenue of research further.

As stated at the beginning of this book, my meta-challenge in this research is to convince fellow computer scientists that thorough historical and philosophical reflections can advance computer science itself. I hope this book has contributed to achieving this goal.

Errata

Errata are available at *www.dijkstrascry.com/turingtales*.

Acknowledgments

I am grateful to several colleagues for commenting on drafts of one or more chapters of the present book, including the following people:

- Kurt De Grave
- anonymous referees of the CACM and POPL
- David Parnas
- Liesbeth De Mol
- Erhard Schüttpelz
- Maarten Bullynck
- Matti Tedre
- Giuseppe Primiero
- Thomas Haigh
- Jan Friso Groote

- Harry Lintsen
- Simone Martini
- David Nofre
- Raymond Boute
- Robert Harper
- Maarten van Emden
- Gerard Alberts
- Bart Demoen
- Eric Berkers
- Raphael Poss

I also thank Rosa Sterkenburg, Giuseppe Primiero, Arthur Fleck, and Raymond Boute for writing parts of the present book. As mentioned in the endnotes, Sterkenburg and Primiero helped with specific parts of Chapter 6, while Fleck and Boute are the sole authors of Chapters 4 and 7, respectively.

A. Jeroen Voeten's "Fundamental Limitations"

Jeroen Voeten's 2001 article, 'On the fundamental limitations of transformational design' [333] covers "transformation systems of designs." The word "design" is used as an umbrella term because it can refer to either a circuit or a computer program, e.g., described in VHDL or expressed in C, respectively.

Voeten formalizes a "transformational design system" and subsequently extends the notion of formal language to an "executable language" by resorting to the familiar Turing-machine concept. The mathematics largely follows the textbook of Harry Lewis & Christos Papadimitriou [226] and is watertight. I thus concur with Voeten's purely theoretical result, which states that

> [C]omplete transformation systems for Turing complete languages do not exist. [333, p.540]

A complete transformation system, as Voeten clarifies, "guarantees that any correct design can in principle be explored using the transformation system" at hand [333, p.533].

Less obvious is Voeten's assertion that

> [M]any design languages used in every-day practice, such as C, VHDL and many flow graph languages, are Turing complete. [333, p.533]

Again, *modeling* such technical artefacts with a Turing-complete language is just fine. However, is Voeten aware that he is modeling in the first place? Does he realize that he has chosen specific abstractions at the expense of others so that his favorite Turing-machine language becomes an 'adequate' modeling language for the technical artefact at hand?

Voeten notes that "incompleteness" of a collection of transformations is "not well understood" and "it is often believed that complete transformation systems can be constructed" in practice [333, p.533]. However, Voeten incorrectly assumes that there is only one way to mathematically model VHDL, C, and corresponding transformation systems; i.e., by resorting to a universal Turing machine. Several mathematical models of VHDL exist, not just one. Moreover, as stressed before, a mathematical model can never be equivalent to the technical artefact at hand: VHDL and C cannot *be* Turing-complete languages because they belong to another category of objects than the category of abstract objects. As a result, Voeten's formal apparatus only allows us to conclude that if we were to model, say, VHDL in the specific way he has done, then there are fundamental limits on what we can do in practice with his model.

Now it should be no surprise that I fundamentally disagree with Voeten's main take-away message:

> In this article, we show, using a formal framework based on the theory of computation, that this is not the case [i.e., complete transformation systems cannot be constructed] if the transformation system is based on an expressive general-purpose design language such as VHDL. [333, p.533]

Voeten's impossibility claims only hold with regard to his Turing-complete modeling language. No absolute incomputability claim can be made about VHDL and C in general. This rectification, then, is yet another specific technical contribution of the present book.

Voeten continues by stating that "only when restrictions are imposed on the design language and correctness relation" a transformation system can be made "complete in theory, but this is expected to result in serious practical problems" [333, p.533]. What Voeten actually means is that if one is to prefer his mathematical model and its underlying abstractions, then the only way to obtain theoretical completeness is by imposing further restrictions in the way he suggests.

Voeten's initial restrictions, spelled out in the main body of his paper, still comply with the fundamental abstractions that he has implicitly made in the beginning of his exposition. For example, Voeten discusses the possibility of stepping away from the Turing-machine model of computation and resorting, à la Meyer & Ritchie, to the more restricted class of primitive recursive functions instead. However, as we have already seen, also the latter modeling approach complies with fundamental abstraction A_I^{pos}.

Fortunately, towards the end of his paper, Voeten does consider the possibility of restricting the class of primitive recursive functions to only those functions that are defined on a finite domain D. In my own words, Voeten peruses breaking down abstraction A_I^{pos}, which I view as a very good sign. Unfortunately, Voeten does not explore this avenue further in his theoretical analysis; instead, he dismisses it on intuitive grounds. In his words:

> We must further realize that in order to make a transformation system complete, we must exploit the finiteness property of the domain. It is easy to prove that this must result in non-universal transformations that are only valid in the context of the chosen finite domain. We then expect the resulting transformation system to be terribly complicated [. . .] but we have no clue how to prove this intuition. [333, p.542].

However, Voeten has proposed *only one possible way* to mathematically capture the finiteness of a physical computation, namely by first abstracting to the class of primitive recursive functions and *only then* imposing a finiteness condition on each function in that class. Dijkstra, a former researcher from Voeten's university, would have approached the same problem very differently. In other words, Voeten's analysis does not address alternative approaches that lie outside the mainstream, Turing-complete, way of thinking.[112]

Endnotes

[1]I wrote most of Chapters 2 and 3 a few years ago and the careful scholar will observe that some statements made in the endnotes exemplify category mistakes. I think it is instructive to keep these mistakes in the first edition of the present book.

[2]Paraphrasing Michael Jackson [193, p. 51].

[3]See my blog post on Michael Mahoney [99].

[4]See McCarthy's recollections [239, p.191] and Herbert Stoyan's historiography [308].

[5]It should also be noted that the Turing machine model of computation has been over sold by computer scientists [35] and historians [99] alike. But only in future work will I attempt to rectify this situation in a more direct manner; e.g., by focusing on the legacy of Alonzo Church and the topic of types in programming languages [237], and e.g., by scrutinizing the relationship between reasoning and computing [116].

It should perhaps also be remarked that the technicalities in the present edition of this book make it less accessible for some fellow historians, which I regret and hope to remedy in the future.

[6]Sources: Copeland [76, p.3], Davis [90, p.147,166], and Hodges [183, p.1994]. For a further discussion, see the present author's paper 'A Turing Tale' [101].

[7]Paraphrasing Parnas, based on private correspondence with the present author in early August 2016.

[8]I have written that "several prominent computer scientists ...

actually believe ..." This statement is based on my own research and on the many examples provided in the writings of Timothy Colburn [72] and Donald MacKenzie [233].

[9]Details are provided in Chapter 6 along with the following definition: a mathematical language is *Turing complete* when it is able to express all partially computable functions [333, p.539].

[10]Three remarks are in order. First, to be more precise, the diagonal argument can be applied only to mathematical objects *if* one wants to maintain the claim that one is using the argument in a mathematical proof.

Second, no harm is done in hijacking the term "computer program" and using it as a synonym for a "mathematical program," but only as long as one does so (a) *consistently* and (b) without claiming that the obtained result is an absolute claim about *technology*.

Third, and for the sociological record only, it should be noted that the reviews received from the CACM are much more elaborate than those from POPL. The CACM reviews convey more how each reviewer thinks and are several pages long. In retrospect, this is perhaps to be expected; the CACM is after all a journal, while POPL is an annual conference with a well-defined scope of research.

[11]Two remarks. First, Strachey was passionate about the lambda calculus rather than Turing machines, but this observation is less relevant for the purpose of the present book. Second, I refer to my oral histories with Naur and Jackson for more coverage on their views [96, 193].

[12]I consistently follow Turner in writing "artefact" (British English) instead of "artifact" (American English) even though the rest of this book is written in American English.

[13]I am particularly grateful to Thomas Haigh for his written and oral feedback on multiple drafts of this chapter, starting in the spring of 2013. I also thank Nancy R. Miller and J.M. Duffin of the University of Pennsylvania Archives and Jos Baeten of the 'Centrum voor Wiskunde & Informatica' for funding my visit to the Archives.

[14]The minutes of that meeting state:

> Bright reported that the Program Committee recommends that the National ACM Lecture be named the Allen [sic] M. Turing Lecture.

> Oettinger moved, seconded by Young that it be so named. Several council members indicated they were not satisfied with this choice.
>
> Juncosa suggested we consider a lecture name that is not that of a person. van Wormer moved, seconded by Juncosa to table the motion.
>
> The vote was: for-15; opposed-5; abstention-2. [1, p.11, original emphasis]

[15]Carr was President of the National Council of the ACM in 1957–59, founding editor of Computing Reviews in 1960–62, and member of the Committee for the Turing Award during the second half of the 1960s. Gorn was committee member on programming languages, a Council Member in 1958–68, and Chairman of the Standards Committee in 1962–68. Oettinger was President of the ACM in 1966-68 [4, 59, 225, 260, 268].

[16]The phrase "a large store containing both numbers and instructions" is akin to the terminology used by Andrew & Kathleen Booth in 1956 [41]. The primary sources listed in the bibliography suggest that the Booths, Carr, Gorn, and several other actors did not use the words "stored program" during the first half of the 1950s.

As Mark Priestley notes in his book, *A Science of Operations* [280], it was the early machines (such as the ASCC and ENIAC) that were described as revolutionary: "It was several years until the first machines based on the stored-program design became operational, and even longer until they were widely available" [280, p.147].

[17]See Bar-Hillel [26, p.7], Locke [230, p.2,3,12,24], and Oettinger [268, p.207].

[18]The reader should bear in mind that although I have extensively studied the "Saul Gorn Papers," the papers do not contain many primary sources of the 1940s and early 1950s. Moreover, I have yet to travel to the Great Lakes in order to study the "Alan J. Perlis Papers" at the Charles Babbage Institute and "Arthur W. Burks Papers" at the Bentley Historical Library. In future research I also aspire to cover the work of Maurice Wilkes and Christopher Strachey. Complementary to all this, I have contributed to a Dutch book *De geest van de computer* [33] in which I discuss the accomplishments of the Dutch computer builders Gerrit Blaauw and Willem van der Poel and the programmers Adriaan van Wijngaarden and Edsger W. Dijkstra. See Chapter 3 in the present book for a translation of part of my Dutch exposition.

[19]A complementary history of programming languages is covered in Donald Knuth's 'The Early Development of Programming Languages' [209, Ch.1]. While Carr and Gorn are not mentioned in the main body of Knuth's account, they are key players in the present chapter.

[20]The words "machine translation" and "automatic language translation" are used interchangeably in this book.

[21]See Akera [4] and Wilkes [342].

In retrospect, it is safer to restrict the claim that "everybody knew each other" to British and American practitioners. Grace Hopper's 1978 recollection illustrates this point:

> I had absolutely no idea of what [the Swiss] Rutishauser, [the German] Zuse, or anyone else was doing. That word had not come over. The only other country that we knew of that was doing anything with the work was Wilkes in England. The other information had not come across. There was little communication, and I think no real communication with Germany until the time of ALGOL, until our first ALGOL group went over there to work with them. I think it's difficult for you in a seminar here to realize a time when there wasn't any ACM, there wasn't any IEEE Computer Society. There was no communication. There was no way to publish papers. [335, p.23]

[22]That "robot," which was of a "certain formal character", could deduce "any legitimate conclusion from a finite set of premises." Weaver adapted McCulloch and Pitts's theorem to mean that "insofar as written language is an expression of logical character," the problem of machine translation is at least solvable in a formal sense [230, p.22].

Neither Weaver nor McCulloch and Pitts explicitly referred to Turing's papers. Moreover, I have given no evidence here to suggest that Weaver was, by 1949, well versed in Turing's theoretical work. Likewise, McCulloch and Pitts's sole, brief, and inadequate reference to a "Turing machine" in one of their concluding remarks might suggest that they were not well versed in Turing's theory of computation either. Their sole reference to a "Turing machine" was, strictly speaking, incorrect. In their words:

> [E]very net, if furnished with a tape, scanners connected to afferents, and suitable efferents to perform the necessary

motor-operations, can compute only such numbers as can a [universal] Turing machine [240].

Nevertheless, an indirect link between Weaver's research agenda on machine translation and Turing's 1936 notion of a universal machine had been established by 1949, even though Turing's work had yet to really surface in both linguistics and computing. For example, no reference was made to Turing, Post, and the like in the comprehensive 1955 book *Machine Translation of Languages* [230].

[23] For a good understanding of the connection between McCulloch & Pitts's 1943 paper and von Neumann's work on the EDVAC, see Akera [5, p.118–119] *and* Burks [54, p.188] in conjunction.

[24] Bar-Hillel was moreover coming to the conclusion that "even machines with learning capabilities [...] will not be able to become fully autonomous, high-quality translators" [26, p.9].

[25] Likewise, no mention of Turing was made in the comprehensive 1971 book *Computer Structures: Readings and Examples* [29]. I thank David Parnas for bringing this book to my attention. Turing's involvement with computer *building* was *popularized* later, by Brian Randell (1973), Andrew Hodges (1983), John Robinson (1994), Martin Davis (2000), George Dyson (2012), and others [89, 117, 182, 285, 292]. David Kahn's 1967 book *The Codebreakers* [199] did not cover much about Turing or von Neumann. I thank Vinton Cerf for bringing this book to my attention.

[26] Contrast this with Dyson's claim that Booth saw von Neumann's computer project at Princeton as the practical implementation of both Babbage's *and* Turing's ideas [117, p.132]. For further scrutiny of Dyson's book *Turing's Cathedral*, I refer to my review [97].

[27] In modern terminology: Most computer designers who considered using a small instruction set did *not* do this in connection with practically realizing a universal Turing machine. An exception in this regard, other than Turing himself, was the successful Dutch computer builder Willem van der Poel, as we shall see in Chapter 3.

[28] Several chapters in Bowden's 1953 book *Faster than Thought* [48] honored Babbage in a similar manner. Chapter 8 on the ACE *also* made a connection between Turing and Womersley's computer building aspirations on the one hand, and Turing's 1936 paper on the other hand. But, as the "Türing Machine" entry on page 414 indicates, the vast majority of computer builders — and practitioners in general — did not read Turing's 1936 paper, let alone grasp it's practical implications

(cf. [98]).

^{29}I thank Thomas Haigh for making this correction.

^{30}So far, my exposition aligns well with one of Akera's main claims, namely that

> [a]side from the relative ease by which a program could be set up on such a system, the major aim of [EDVAC's] design was to reduce the amount of hardware by severely limiting the machine's parallelism. [5, p.115–116]

^{31}A discussion of von Neumann's own writings lies outside the scope of this book but see Haigh et al. [170]. I speculate that von Neumann knew that a store was *not* a necessary condition (nor a sufficient condition) for realizing a practical version of a universal Turing machine (cf. endnote 14 in [98, p.202]). Or, as Allan Olley befittingly puts it, "the stored program [computer] is not the only way to achieve [what was later called] a Turing complete machine" [269]. The terms "Turing" and "completeness" were in use by 1965 (cf. J.G. Sanderson [298]).

Von Neumann was in a position to see that the universal Turing machine was a mathematical model of his IAS computer, the EDVAC, *and* the ENIAC. There are no primary sources supporting the popular belief that he thought he was building the first practical realization of a universal Turing machine. Priestley has drawn very similar conclusions in his book [280, p. 139, 147–148]. More important then is Akera's observation that von Neumann was embarking on a "formal theory of automata" — a theory which pertained to "machines capable of modifying their own programs" [5, p.119]. And, just to be clear, the ability to modify its own programs is *not* a prerequisite to practically realizing a universal Turing machine.

^{32}An erasable store was *not* a necessary condition (nor a sufficient condition) for realizing a practical version of a universal Turing machine — a fact known to the mathematical logician Hao Wang in 1954 who *did* ponder about these theoretical issues [98, p.21].

^{33}From a *sociological* perspective, it is interesting to compare and contrast Hopper's words with the following words made by Andrew Appel in 2012:

> [Appel, 27 minutes and 44 seconds into his talk:] The machines we use today are von-Neumann machines. And

von Neumann was quite clearly influenced by Turing in building universal machines with a finite-state control and infinite tape. Really the only difference is that the infinite tape is random access rather than linear access. So von-Neumann machines are what we now call computers. The Harvard architecture in which the program is not stored in the memory of the computer; well, we saw a quotation from Howard Aiken [in the previous talk by Martin Davis [91] — a quote which Davis repeatedly uses to ridicule Aiken and which I object to in my previous book [98, Ch.8] and in my article for the Communications of the ACM [101]]. By the 90s, when you called something a Harvard architecture, you meant that the program was stored in a memory but in a read-only memory because nobody would even imagine that you could have a computer program that wasn't even representable as bits. [14]

Both Hopper and Appel were using the historical actor "Howard Aiken" and "history" in general to make a case for their own research aspirations. I view Hopper as a pluralist; in her 1978 keynote address, she repeatedly encouraged a multitude of programming styles as opposed to "trying to force" every practitioner into "the pattern of the mathematical logician":

> [Hopper:] I'm hoping that the development of the micro-computer will bring us back to reality and to recognizing that we have a large variety of people out there who want to solve problems, some of whom are symbol-oriented, some of whom are word-oriented, and that they are going to need different kinds of languages rather than trying to force them all into the pattern of the *mathematical logician*. A lot of them are not. [335, p.11, my emphasis]

I view Appel as a unifier; that is, as a theoretically-inclined researcher who seeks unifying principles to further his research. Appel's 2012 address was all about the importance of mathematical logic in computing and thus, unsurprisingly, about Turing and von Neumann's allegedly important roles in the history of the computer.

[34]An anonymous reviewer is quite right in noting that the words "In my eyes" run the risk of judging Carr's work in anticipation of later ways of thinking. To rectify this, I include this note as a disclaimer (which has another, intended, effect than simply removing the quoted words from the narrative).

[35]While it was common practice to refer to the "universal language of mathematics" and while it was customary to associate a language with machine code, as in the words "machine language," it was in 1954 still uncommon to describe the pseudo code of a computer as a "language." Brown and Car *did* consistently do this in their joint paper. Backus and Herrick came in a close second at that same symposium. And, as Nofre et al. point out, the words "programming language" in fact only entered the computing arena during the mid-1950s [261].

Nofre et al. state that already by the early 1950s the translation from a mathematical language to machine language had "become a central metaphor used to make sense of the activity of programming."

[36]See Gorn [158, p.74–75].

Nofre et al. describe the origins of the idea of a universal code, going back to C.W. Adams of MIT in 1951.

[37]Note that Gorn made this implication at a time when computer storage was limited.

[38]As the 1950s progressed, Carr and Gorn refined their expositions. For example, "compilers" did "not absolutely require their machines to possess common storage of instructions and the data they process" but they were "considerably *simpler* when their machines" did have this property [160, p.254, my emphasis].

[39]This fourth requirement, by itself, already indicates quite strongly that Carr and Gorn were not pursuing logical minimalism à la Turing and van der Poel (cf. De Mol & Bullynck [247]).

[40]These properties were sufficient (but *not* necessary) for realizing a practical version of a universal Turing machine. Note, again, that I am making the theoretical connection here *instead of* the historical actors. Carr and Gorn had the knowledge to defer this conclusion themselves, but I conjecture that they were *not* reasoning along these lines *at all*, during the first half of the 1950s (and possibly throughout their whole careers).

[41]Machines that were not based on the principle of common storage of numbers and instructions were instructed differently than those that did possess loop control. The Swiss numerical analyst Heinz Rutishauser, for example, used (a modification of) Zuse's Z4 machine through part of the 1950s, a machine that did not possess loop control. Although more research on Rutishauser is required, it is already interesting to note here

that only by 1963 did he see the need to program recursively in his field of numerical analysis [297] (see also [94, 98]).

[42]That is, in the finitist terms set forth by "intuitionist mathematicians such as Kronecker, Brouwer, and Weyl" [159].

[43]Some of Gorn's 1955 insights were later published in his 1957 paper [160].

[44]For Gorn, infinite memory represented the simplest case of analysis. This was similar to Curry's reasoning (cf. [77] [314, p.65]) and unlike van der Poel's insistence to view results from mathematical logic solely in the setting of finite machines (cf. [100, 276]).

[45]Besides Gorn in 1956, also Berkeley in 1958 discussed the importance of machines that have "loop control," and hence "think," compared to the "non-thinking machines" [32, p.5]. Contrast these observations with Akera's claim that the historiographical interest in the "stored-program concept" really only started around 1964 with the emergence of theoretical computer science [5, p.120].

[46]Notice that Gorn excluded the "older machines," such as the ENIAC. He was implicitly referring to all loop controlled computers and to the principle of a common store in particular. According to Gorn, those machines provided the flexibility for the development of a universal code.

[47]Moreover, note that Gorn's recognition was, strictly speaking, incorrect. We know, today, that also many of the "older machines" can be viewed as "Turing universal" as well. For example, it suffices to "wire in" the program of a universal Turing machine and then represent any (to-be-simulated) Turing machine program, along with working store, on punched paper tape — tape which serves both as input and output for the calculation at hand. It is *not* a theoretical obstacle that the punched holes cannot be removed (cf. [98, p.22]).

An anonymous reviewer has commented that

> [E]ven if ENIAC was later identified as being Turing complete, the important historical thing is that the first machines to be thought of as Turing complete, and to inspire the integration of Turing's work into the growing field of computer science, were modeled on EDVAC.

This assessment by the reviewer seems fair with regard to Gorn's career.

I have not given evidence to suggest that the claim holds in its general form; i.e., that it holds for other actors like von Neumann, van der Poel, Turing, and Wang. I would not be surprised if von Neumann did connect the ENIAC to Turing's 1936 universal machine around 1945, thereby (partially) contradicting the reviewer's comment in its general form.

In this regard I also present Priestley's words which I think nicely summarize a complementary part (of the total discussion):

> None of the writings originating from the Moore School group mention a logical or theoretical rationale for the introduction of the stored-program concept. It remains a possibility, of course, that the idea was suggested by von Neumann's knowledge of Turing's work, but even if that was so, its inclusion in the design was justified by practical, not theoretical, arguments. A logical pedigree for the idea would not, on its own, have been sufficient to ensure its incorporation in the design without a detailed examination of its engineering implications. [280, p. 138-139]

I conjecture that von Neumann knew that *both* the ENIAC and the EDVAC were "Turing universal" and he knew that this insight had no immediate practical relevance on his computer-*building* aspirations.

(In retrospect: Chapters 5 and 6 in this book show that the previous reasoning contains category mistakes. Specifically, the ENIAC and the EDVAC can at best be *modeled* with a universal Turing machine. They cannot *be* Turing universal.)

[48]Thanks in no small part to the writings of Burks (cf. [98, p.25]).

[49]Gorn continued by stating that the as yet ill-defined, universal command language is "probably, in terms of mathematical logic" equivalent with "the class of expressions obtainable by 'general recursive definitions'." In the early 1960s, the recursive-function theorist Henry Gordon Rice substantiated Gorn's hunch by making an explicit connection between ALGOL 60's computational power and the class of general recursive functions [98, Ch.8]. Important to note here is that, once again, the *theoretical* insights came *after* programming *practice*.

[50]See Carr [61, p.230, 259].

Post's work was, together with Davis's 1958 book [86], also largely responsible for the advent of "decision procedures" [61, p.223, 225] in automatic programming, automata theory, and mathematical linguistics in general. It was researchers in computational linguistics (such as Chomsky, Bar-Hillel, and especially Micha Perles and Eli Shamir) and soon-to-be-called theoretical computer scientists (such as Rabin, Scott, Ginsburg, and Rice) who demonstrated a thorough reception of Turing's work and especially Post's "correspondence problem" during the late 1950s and 1960s (see e.g., [27, 151, 283]).

[51]Note that some of these insights were already several years old. I am discussing Carr's 1959 work here because I have yet to come across much of his earlier writings.

[52]Turing's 1936 paper and his follow-up correction [320, 321] were reprinted in Appendix One of the proceedings [157].

[53]I conjecture that Turing never viewed his 1936 paper from this angle — i.e., that he was not (and did not want to be) a space cadet — and that Booth and other space cadets did not completely grasp Turing's work. On the one hand, of course, *that* is exactly where the force of Turing's theory lies; it helped some researchers — and, more importantly, practitioners working in or with industry — to see the wood for the trees in unanticipated ways. On the other hand, many practitioners were not really helped by Turing's work: there is no evidence suggesting that if Turing had not existed, that these practitioners would then have been doing something else. Many advocates of various programming languages knew that these languages were equivalent. Several languages were implemented by translating into another competing language. The real issue was their suitability for human use. (I thank Parnas for discussing this matter with me.) Having said *that*, I now refer to Robert Bemer's experience with equivalent languages. In 1957 he wrote:

> Although the ultimate in language does not exist yet, we can console ourselves meanwhile with compatible (as against common) language. There is much current evidence that existing algebraic languages are all mappable into one another by pre-processors. [30, p.115][261]

Bemer's words make me conclude that some researchers were in need of a solid, theoretical, unifying principle, such as the universal Turing machine or the universal programming language ALGOL. Others were satisfied with experimental evidence. My conclusion here is merely another way of expressing my respect for the *multitude* of personal styles in tackling technical problems, a topic that I have discussed at length with Peter Naur [96]. Unsurprisingly, it was the *theoretically-*

inclined space cadets (Carr, Gorn, and Perlis) who talked about some implications of Turing's work during the 1950s.

[54]In line with Nofre et al. [261], I am introducing the word "metaphor" here instead of the historical actors. Nofre et al. elegantly discuss the origins of the metaphor, going back to postwar cybernetics and George Stibitz's role in 1947 in this regard. I take gentle issue with their supposition that the language metaphor had become a central metaphor in the computer programming arena by the early 1950s. Carr and his space cadets were few in number, exactly because they took the language metaphor seriously.

[55]Regardless of whether Gill was the first to articulate the onion-skin metaphor, it should at least not go unnoticed here that hierarchical design would become a hot research topic in later years. In what I take as support for my narrative, Priestley's 'The Invention of Programming Languages' [280, Ch.8] suggests that (i) men like Turing (1951), Wilkes (1952), and others understood the idea of "an interpretive routine enabling one machine to simulate another" and that (ii) in *later* years, Booth, Gill, and others associated the universal Turing machine concept with interpreters for high level programming languages [280, p.191-192].

[56]Gorn's technical reports from the 1960s and onwards were all about: Chomsky — Algol — Turing — Gödel — Wang — Curry — Carnap — Quine — Rosenbloom — . . . and so on.

[57]There are other examples. Philip R. Bagley, for instance, connected the Universal Computer-Oriented Language (UNCOL) to Turing's work. I take his understanding of Turing's work to be weak (and without much practical significance to him), as his words from 1960 partially indicate:

> If UNCOL can express the basic computing steps required by a [**universal**] Turing machine, then since all present-day computers are equivalent to [finite realizations of **universal**] Turing machines, it can lay claim to being "universal" from the standpoint of computation alone. [. . .] The limitation of UNCOL that we are really concerned with is whether or not we can achieve tolerable efficiency of execution [. . .] [23, p.21] [See also [22]]

[58]Weaver supported the idea of an indirect route: translate from the source language into an as yet undiscovered *universal* language and then translate from that language into the target language [230, p.2,23]. Likewise, Edsger Dijkstra (and others) supported a two-step translation

of the universal programming language ALGOL 60. First, translate the ALGOL 60 program into an intermediate *machine-independent* object language. Second, process the obtained object program by an interpreter (which is written in the machine code of the target computer) [98, p.68]. Also UNCOL was based on this principle, which was

> the design of a *universal*, intermediate language, independent of specific hardware but similar in character to machine languages, as a bridge over the growing gap between problem oriented languages (POLs) and machine languages (MLs). [326, p.2, my emphasis]

[59]In other words: not only did the space cadets associate the machine with a "machine language" and the pseudo code with a "programming language," they also began to view a programming language as a machine. Both metaphors together led to hierarchical decompositions of software and hardware in which each level was characterized as both a machine and a language.

[60]Already in 1962, Dijkstra wrote: "machine and language are two faces of the same coin" [108, p.237]. Likewise, Adriaan van Wijngaarden wrote: "we rather see the language as a machine" [337, p.18].

[61]Nofre et al. furthermore state that "two prominent senses of the notion of universality emerged" during the 1950s; namely, the idea of "machine independence" and "a sense connecting directly with the universality of the notations of science, and in particular mathematical notation." Two more prominent senses of universality can now be mentioned, namely the idea of a universal Turing machine and a sense connected directly or indirectly with the universality of Weaver's interlanguage.

[62]I thank Parnas for discussing this matter with me. See also Chapters 3 and 6 in this book.

[63]ARCO: "Automatische Relais Calculator voor Optische berekeningen" (Automatic Relay Calculator for Optical calculations)

PTT: "Posterijen, Telegrafie en Telefonie" (Mail, Telegraphy, and Telephony)

[64]See Kranakis [212, p.76–77] and Rooijendijk [293, p.154–159].

Nat Lab: "Natuurkundig Laboratorium" (Physics Laboratory)

PETER: "Philips Experimentele Tweetallige Electronische Rekenma-chine" (Philips Experimental Bi-lingual Electronic Calculating machine)

[65]In other work I argue that some of these Dutch advancements had an impact on the international scene during the 1960s and 1970s [33].

[66]This theme has already been addressed in the Dutch literature by Alberts & de Beer (who write about "dream machines" [7, p.101]), by van den Bogaard (who writes about "Dijkstra: programming without computers" [39, p.137]) and, more recently, by Daylight (who connects the Amsterdam school of machine-independent programming with Blaauw's later work on IBM's "compatible machines"; see [33]).

[67]Logic:

- Hilbert und Ackermann. Grundsüge der theoretischen Logik. Springer, 1938 en Dover, New York 1946.

Modern discontinuous machines. Fundamentals:

- A. Turing. On Computable Numbers. Proceedings of the London Mathematical Society, 1937.

- C. Shannon. A Symbolic Analysis of Relay Circuits. Electrical Engineering, 1938.

- A.W. Burks, H.H. Goldstine & J. von Neumann. Preliminary discussion of the Logical Design of an Electronic Computing Instrument.

- Goldstine & Neumann. Planning and Coding of Problems for an Electronic Computing Instrument. Vol. I & II.

- [Referring to the latter two:] Both reports appeared in: The Institute for Advanced Study. Princeton, N.J. 1946 – '48

Continuous machines:

- Francis J. Murray. The Theory of Mathematical Machines. King's Crown Press. New York, 1948.

[68]PTERA: "PTT Elektronische Reken Automaat" (PTT Electronic Calculating Automaton)

For a more detailed account of Kosten, see Kranakis [212, p.70–72] who says that before van der Poel's arrival, Kosten had already shifted the attention of his research group to developing a general-purpose computer.

[69]ARRA: "Automatische Relais Rekenmachine Amsterdam" (Automatic Relay Calculating machine Amsterdam)

See Alberts & de Beer [7], Daylight [95], and Kranakis [212, p.64].

[70]See Bloem [37], Daylight [95], and Aiken's biography [71].

[71]See Alberts & de Beer [7, p.110], Bloem [37], and Daylight [95].

In later years, researchers did distinguish between the ARRA I and ARRA II; see van Donselaar [115].

For the Amsterdam machines that were built after Blaauw's departure (e.g., ARMAC, X1), see Alberts & de Beer [7] and Kranakis [212].

FERTA: "Fokker Electronische Rekenmachine Type ARRA"

ARMAC: "Automatische Electronische Rekenmachine Amsterdam"

[72]See Alberts & de Beer [7], Bloem [37], Daylight [95], and Kranakis [212].

[73]I borrow the word "contract" only once from Alberts & de Beer [7] since this comes from a secondary source, i.e., from Dijkstra's own recollections [114].

[74]I presume the term "Cycle program" overlaps with what John Carr and Saul Gorn called a "loop controlled" program; see Chapter 2.

[75]ZERO: "Zeer Eenvoudig Reken Orgaan" (Very Simple Calculating Machine)

See Daylight [93] and van der Poel [275].

[76]I borrow the term "logical minimalism" from De Mol & Bullynck [247].

The technical content of my exposition comes from van der Poel [275] and Verhagen [331, p.31].

[77]Quoted from van der Poel [276, p.18-19].

ZEBRA: "Zeer Eenvoudig Binair Rekenapparaat" (Very Simple Binary Calculating machine)

[78]See Kranakis [212, p.74] and Bogaard [39].

[79]For example, $\neg a$ can be rewritten as $a \uparrow a$ and $a \vee b$ can be rewritten as $(a \uparrow a) \uparrow (b \uparrow b)$.

[80]For Gorn, infinite memory represented the simplest case of analysis. This was somewhat similar to Haskell Curry's reasoning (cf. [77] [314, p.65]) and was unlike van der Poel's insistence on viewing results from mathematical logic solely in the setting of finite machines (cf. [100, 276]). The following words by J.W. Waite, Jr. at a 1956 symposium on automatic programming are most instructive:

> One of Professor Curry's most significant comments [reference [77]] on programming is: "You ask why I assumed an *infinite memory* in my analysis. It is the usual practice of the mathematician to proceed from the simple to the more complex. In this case, the infinite memory represented the simplest case. For finite memories, man-made constraints lead to more complex considerations. I have only attempted to show the way. What I see evolving is a new calculus of logic, that is[,] programming logic. It is the responsibility of those in the field to develop this new calculus." [314, p.65, my emphasis]

I thank Liesbeth De Mol for bringing this source to my attention. For more on Curry, see De Mol [248].

[81]In 1961, van der Poel gave his programming style a name: "trickology" [39, p.137]. In his words:

> The concept of micro-programming and the practice of devising tricks to do the more complicated composite action is so interwoven in ZEBRA that the volume of knowledge of these tricks has been given a special name: *trickology*. Without this knowledge of trickology and the standard programs based on it, ZEBRA would be a useless machine. [277, p.274, original emphasis]

[82]See Wilkes [342, p.31], Akera [4], and Alberts & de Beer [7, p.121].

[83]UNIX is a registered trademark of The Open Group.

[84]See Turner [324]. `Miranda` is a trademark of Research Software Ltd.

[85]One can argue that the `C` standard does not describe *one* `C` computer programming language but *several*. I look forward to contemplating precisely this kind of remark in future research and hopefully with both philosophers and 'POPL insiders.'

[86]Strictly speaking, I should not write a "program text" but rather a "program text printed on paper" or a "program text projected onto a screen." The latter two objects are technical artefacts while one could argue that a plain "program text" is an abstract object.

[87]I thank a reviewer for making the following observation: the Timed Automata model is another example that does not rely on finite abstractions and yet is complete and decidable.

[88]Michelle Obama said this at the Democratic National Convention in the summer of 2016. My source is a youtube video: *www.youtube.com/watch?v=mu_hCThhzWU*

[89]I also mention the new ontological category of "liminal artefacts" as a viable alternative [85, Ch.2] and look forward to applying Roman Frigg's "literary fiction" approach [145] towards modeling in future work. I am after all, as this book illustrates, sympathetic to the viewpoint that *computer* programs are neither purely mathematical nor physical.

[90]The word "we" refers to my former student, Rosa Sterkenburg, and myself. Rosa is a co-author of the present introduction (Section 6.1) but carries no responsibility for the rest of this chapter.

[91]We have tried to follow Brian Cantwell Smith's thoughts in the main text and use this note to highlight a necessary distinction that has to be made between a computer program and a mathematical program: We can mathematically prove that *a mathematical counterpart* of our computer program respects the mathematical model (i.e., formal specification). However, in the real world there exist red apples as well. So our formal specification is not perfect and, therefore, neither is our mathematical program and nor is our computer program.

[92]I could equally well consider some related issues — such as, loosely speaking, (1) an unbounded number of memory locations that can be addressed, and (2) an unbounded length of programs that can be compiled — but choose not to do so here.

[93]Note that stating the existence of abstract objects does not commit

one to believe in a platonic realm. I thank Giuseppe Primiero for sharing this observation with me.

[94]Spelling these trivialities out might give the false impression that 'POPL insiders' are completely unaware of, or would strongly disagree with, these clarifications. I don't think this is the case. I suspect 'POPL insiders' simply do not spend much time scrutinizing the *informal* aspects of the topic at hand, at least not in published writings.

[95]The following four subsections are partly based on collaboration with philosopher Giuseppe Primiero. All rights are reserved by the present author.

[96]A mathematical language is Turing complete when it is able to express all partially computable functions [333, p.539].

[97]This view has been expressed by referees of the CACM. In this particular case I have not cited but paraphrased the referees. I suspect the CACM referees are *not* programming language experts.

Two more remarks follows. First, I take no issue with the claim that the lambda calculus is a mathematical programming language as long as one does not equate the latter with a computer programming language. Second, a "computer programming language" is often abbreviated to "programming language" in many of my prior writings (and in several received review comments). In the present chapter, however, I try to consistently use either the adjective "computer" or "mathematical" in front of the phrase "programming language."

[98]Raymond Boute's contribution in Chapter 7 reveals another reality: even the *function* concept has no standard definition today. Specifically, Boute observes that, unlike most analysis/calculus texts, many *introduction to proof* texts do not define the *function* concept properly and fully generally.

[99]I anticipate that more clarity will be obtained in the near future by software scholars, i.e., philosophically-inclined researchers who work *across* academic boundaries. For example, Nurbay Irmak has investigated the similarities and differences between music and software [190]. His results have partially inspired Raymond Turner to write his insightful paper [325] which provides other reasons why computer programming languages, such as L, are technical artefacts.

[100]Strictly speaking, the depicted symbols constitute an $\mathcal{L}$-program text printed on paper and that representation (i.e., technical artefact), in

turn, denotes an $\mathcal{L}$-program (i.e., a mathematical object).

[101]I am grateful to an anonymous referee for pointing this out. In a previous version of my writing, I had blindly (and ironically) followed van Wijngaarden by conflating actual computers and mathematical models. The whole point of my exposition is to disentangle both concepts.

[102]Cited in my book *The Dawn of Software Engineering: from Turing to Dijkstra* [98, p.62]. The source is one of Dijkstra's visuals which he used at a conference in Geneva in 1973.

[103]A similar observation was already made by Michael Jackson as follows: "To turn Dijkstra's famous observation on its head, we should say that in a cyber-physical system, where the computer's purpose is to exercise control over the material world, formal reasoning can demonstrate the presence of errors, but never its absence." [193, p. 67]

[104]Nevertheless, I am all for formal methods in industry. I see my colleagues test and debug on a daily basis and realize that so much is to be gained from formal methods.

[105]See Fernandez [124, p.4], Harel [174, p.71], Lee [223, p.86], and Shapiro [301, p.198].

[106]An ordered pair is normally written as (x, y). A few texts use $\langle x, y \rangle$, which is a waste of symbols. In fact, one can even write x, y, reserving parentheses for emphasis or disambiguation.

[107]Instead of notations like $\{x_i\}$, rightly criticized by Halmos, a sequence is simply written as x, whereas x_i is its ith element. As deplored in [224], bad habits such as writing $x[i]$ to mean x remain common.

[108]Technical pitfalls of quote marks are discussed by Curry. A methodological pitfall is getting bogged down in metamathematical issues. Since none of the other texts in this bibliography fall victim to using quote marks, this issue need not concern us.

[109]A comprehensive overview of my system later appeared in Science of Computer Programming [104] and I cannot guarantee that my article is free of category mistakes.

[110]Books worth reading in this regard are Matti Tedre's *The Science of Computing: Shaping a Discipline* [317] and Gilles Dowek's *Computation, Proof, Machine: Mathematics Enters a New Age* [116].

[111]Strictly speaking, the term "program text" is insufficient, and should be written as the "program text written on paper" or the "program text projected onto a screen." I have refrained myself from making these distinctions explicit as well throughout the present book (although I did do so in my original, rejected POPL paper).

[112]I share the view that "Turing completeness" is a shabby criterion for comparing the expressive power of mathematical programming languages. Defending this view is, however, not the main purpose of this book.

Bibliography

[1] *ACM Council Meeting* (1965). Available from the "Saul Gorn Papers", the University of Pennsylvania Archives (unprocessed collection).

[2] *ACM Council Meeting* (1966). Available from the "Saul Gorn Papers", the University of Pennsylvania Archives (unprocessed collection).

[3] *ACM Council Meeting* (1966). Available from the "Saul Gorn Papers", the University of Pennsylvania Archives (unprocessed collection).

[4] A. Akera. "Anthony Oettinger interview: January 10–11, 2006". In: *ACM Oral History interviews*. 2006.

[5] A. Akera. *Calculating a Natural World: Scientists, Engineers, and Computers During the Rise of U.S. Cold War Research*. Cambridge, Massachusetts, USA: MIT Press, 2007.

[6] G. Alberts. "Jaren van berekening". PhD thesis. Universiteit van Amsterdam, 1998.

[7] G. Alberts and H.T. de Beer. "De AERA. Gedroomde machines en de praktijk van het rekenwerk aan het Mathematisch Centrum te Amsterdam". In: *Studium* 2 (2008), pp. 101–127.

[8] G. Alberts and E.G. Daylight. "Universality versus Locality: the Amsterdam Style of ALGOL Implementation". In: *IEEE Annals of the History of Computing* 4 (2014), pp. 52–63.

[9] P. Alevoor, P. Saeda, and K. Kapoor. "On the decidability and matching issues for regex languages". In: *International Conference on Advances in Computing*. Ed. by M.A. Kumar, R. Selvarani, and T.V.S. Kumar. Springer Verlag, 2012, pp. 137–145.

[10] F.L. Alt. *Electronic Digital Computers: Their Use in Science and Engineering*. New York, NY, USA: Academic Press, 1958.

[11] F.L. Alt. "Archaeology of Computers — Reminiscences, 1945–1947". In: *Communications of the ACM* 15.7 (1972), pp. 693–694.

[12] T.M. Apostol. *Calculus, Vol. I (2nd. ed.)* Wiley, 1967.

[13]   A. Appel. The science of deep specification (http://deepspec.org/research/). Accessed on September 25th, 2016.

[14]   A. Appel. *Turing, Gödel, and Church at Princeton in the 1930s*. YouTube: www.youtube.com/watch?v=kO-8RteMwfw. Presentation at the Turing Centennial Celebration at Princeton, 10–12 May 2012.

[15]   V.I. Arnold. *On Teaching Mathematics*. This is an extended text of the address at the discussion on teaching of mathematics in Palais de Decouverte in Paris on 7 March 1997 (http://pauli.uni-muenster.de/~munsteg/arnold.html). 1997.

[16]   J. Backus. "Can programming be liberated from the von Neumann style?" In: *Communications of the ACM* 21.8 (1978), pp. 613–641.

[17]   J. Backus. "The History of FORTRAN I, II, and III". In: *ACM SIGPLAN Notices* 13 (1978), pp. 165–180.

[18]   J. Backus, J. Williams, and E. Wimmers. "An introduction to the programming language FL". In: *Research Topics in Functional Programming*. Ed. by D.A. Turner. Addison-Wesley, 1990, pp. 219–247.

[19]   J.W. Backus. "The syntax and semantics of the proposed international algebraic language of the Zürich ACM-GAMM Conference". In: *IFIP Congress*. UNESCO, Paris. 1959, pp. 120–125.

[20]   J.W. Backus, F.L. Bauer, J. Green, C. Katz, J. McCarthy, A.J. Perlis, H. Rutishauser, K. Samelson, B. Vauquois, J.H. Wegstein, A. van Wijngaarden, and M. Woodger. "Report on the algorithmic language ALGOL 60". In: *Communications of the ACM* 3.5 (1960). Editor: P. Naur, pp. 299–314.

[21]   J.W. Backus, F.L. Bauer, J. Green, C. Katz, J. McCarthy, A.J. Perlis, H. Rutishauser, K. Samelson, B. Vauquois, J.H. Wegstein, A. van Wijngaarden, M. Woodger, and P. Naur. "Revised report on the algorithmic language ALGOL 60". In: *Communications of the ACM* 6.1 (1963), pp. 1–17.

[22]   P.R. Bagley. *Letter to Mort Bernstein*. From the Charles Babbage Institute collections. Thanks to David Nofre for giving me a copy of this letter. 1960.

[23]   P.R. Bagley. *Letter to the SHARE UNCOL Committee and other interested parties, 26 May 1960*. From the Charles Babbage Institute collections. Thanks to David Nofre for giving me a copy of this letter. 1960.

[24]   T.P. Baker and A.C. Fleck. "A Note on Pascal Scopes". In: *Pascal News* 17 (1980), p. 62.

[25] T.P. Baker and A.C. Fleck. "Does Scope = Block in Pascal?" In: *Pascal News* 17 (1980), pp. 60–61.

[26] Y. Bar-Hillel. *Language and Information: Selected Essays on their Theory and Application*. Reading, Massachusetts and Jerusalem, Israel: Addison-Wesley Publishing Company, Inc. and the Jerusalem Academic Press Ltd, 1964.

[27] Y. Bar-Hillel, M. Perles, and E. Shamir. "On formal properties of simple phrase structure grammars". In: *Zeitschrift für Phonetik, Sprachwissenschaft und Kommunikationsforschung* 14 (1961), pp. 143–172.

[28] R.G. Bartle. *The Elements of Real Analysis*. Wiley, 1964.

[29] C. Gordon Bell and A. Newell. *Computer Structures: Readings and Examples*. New York, USA: McGraw Hill, 1971.

[30] R.W. Bemer. "The Status of Automatic Programming for Scientific Problems". In: *The Fourth Annual Computer Applications Symposium, October 24–25, 1957*. Ed. by F.C. Bock. Armour Research Foundation of Illinois Institute of Technology, 1958, pp. 107–117.

[31] E.C. Berkeley. *Giant Brains, or Machines That Think*. New York: Wiley, 1949.

[32] E.C. Berkeley. *Memorandum for the Association for Computing Machinery — Committee on the Social Responsibilities of Computer Scientists*. Tech. rep. 1958. Available from the "Saul Gorn Papers" from the University of Pennsylvania Archives (unprocessed collection): UPT 50 G671 Box 3.

[33] E. Berkers and E.G. Daylight. *De geest van de computer: Een geschiedenis van software in Nederland*. ISBN 978-90-5345-504-3. Uitgeverij Matrijs, 2016.

[34] R. Bird and O. de Moor. *Algebra of Programming*. Prentice Hall, 1997.

[35] G.E. Blelloch and R. Harper. "λ-Calculus: The Other Turing Machine". In: *CMU CSD Fiftieth Anniversary volume*. 2015.

[36] E. Bloch. *Proofs and Fundamentals*. Springer, 2011.

[37] J. Bloem. "Gerrit Blaauw: Van 'rekenmachines' die het doen naar computerarchitectuur". In: *Informatie* (2006).

[38] D.G. Bobrow and B. Raphael. "A comparison of list-processing languages: including a detailed comparison of COMIT, IPL-V, LISP 1.5, and SLIP". In: *Communications of the ACM* (1964), pp. 231–240.

[39] A. van den Bogaard. "Stijlen van Programmeren 1952–1972". In: *Studium* 2 (2008), pp. 128–144.

[40] C. Böhm and G. Jacopini. "Flow diagrams, Turing machines, and languages with only two formation rules". In: *Communications of the ACM* 9.5 (May 1966), pp. 366–371.

[41] A.D. Booth and K.H.V. Booth. *Automatic Digital Calculators*. second. London, UK: Butterworths Scientific Publications, 1956.

[42] N. Bourbaki. *Théorie des ensembles*. Hermann & c^{ie}, 1954.

[43] R. Boute. "Concrete Generic Functionals". In: *Generic Programming*. Ed. by Jeremy Gibbons and Johan Jeuring. Kluwer, 2003, pp. 89–119.

[44] R. Boute. *Rekindling critical thinking: heeding major errors in current* Introduction to Proof *type textbooks*. http://www.funmath .be/CriTnk.pdf. Contributed paper session, MAA Mathfest 2015. 2015.

[45] R. Boute. "Why mathematics needs engineering". In: *Journal of Logical and Algebraic Methods in Programming* 85.5, part 2 (2016). http://dx.doi.org/10.1016/j.jlamp.2016.01.001, pp. 867–878.

[46] R.T. Boute. "System Semantics and Formal Circuit Description". In: *IEEE Transactions on Circuits and Systems* CAS-33.12 (1986), pp. 1219–1231.

[47] R.T. Boute. "Systems Semantics: Principles, Applications, and Implementation". In: *ACM Transactions on Programming Languages and Systems* 10.1 (1988), pp. 118–155.

[48] B.W. Bowden, ed. *Faster Than Thought: A Symposium on Digital Computing Machines*. London: Sir Isaac Pitman & Sons, Ltd., 1953.

[49] J.M. Boyle and A.A. Grau. "An algorithmic semantics for ALGOL 60 identifier denotation". In: *Journal of the ACM* 17 (1970), pp. 361–382.

[50] J. Brown and J.W. Carr III. "Automatic Programming and its Development on the MIDAC". In: *Symposium on Automatic Programming for Digital Computers*. Office of Naval Research, Department of the Navy. Washington D.C., 1954, pp. 84–97.

[51] N.G. de Bruijn. *Verslag inzake onderzoek betreffende electronische en electrische rekenapparatuur over het cursusjaar 1947/48*. Tech. rep. Delft, 1948.

[52] T. Budd. *A Little Smalltalk*. Addison-Wesley, 1987.

[53] M. Bullynck. "Programming primes (1968-1976)". In: *History and Philosophy of Logic* 36 (2015), pp. 229–241.

[54] A.W. Burks. "Turing's Theory of Infinite Computing Machines (1936–1937) and its Relation to the Invention of Finite Electronic Computers (1939–1949)". In: *Theory and Practical Issues on Cellular*

Automata. Ed. by S. Bandini and T. Worsch. London Berlin Heidelberg: Springer, 2001, pp. 179–197.

[55] A.W. Burks. "The invention of the universal electronic computer—how the Electronic Computer Revolution began". In: *Future Generation Computer Systems* 18 (2002), pp. 871–892.

[56] M. Campbell-Kelly. "Alan Turing's Other Universal Machine: Reflections on the Turing ACE computer and its influence". In: *Communications of the ACM* 55.7 (2012), pp. 31–33.

[57] M. Campbell-Kelly and W. Aspray. *Computer: A History of the Information Machine*. New York, NY, USA: Basic Books, 1996.

[58] C. Câmpeanu, K. Salomaa, and S. Yu. "A formal study of practical regular expressions". In: *International Journal of Foundations of Computer Science* 14 (2003), pp. 1007–1018.

[59] J.W. Carr. *Inaugural Presidential Address*. Presented at the meeting of the Association. 1956.

[60] J.W. Carr. *Computing Programming and Artificial Intelligence*. Ann Arbor, University of Michigan. An intensive course for practicing scientists and engineers: lectures given at the University of Michigan. 1958.

[61] J.W. Carr. "Programming and Coding". In: *Handbook of Automation, Computation, and Control: Computers and Data Processing*. Ed. by E.M. Grabbe, S. Ramo, and D.E. Wooldridge. Vol. 2. New York: John Wiley and Sons, Inc., 1959. Chap. 2.

[62] G. Chartrand, A. Polimeni, and P. Zhang. *Mathematical Proofs: A Transition to Advanced Mathematics (3rd. ed.)* Pearson, 2012.

[63] S. Chaudhuri, A. Farzan, and Z. Kincaid. "Consistency Analysis of Decision-Making Programs". In: *Proceedings of the 41st ACM SIGPLAN-SIGACT Symposium on Principles of Programming Languages*. 2014, pp. 555–567.

[64] S. Chaudhuri, S. Gulwani, and R. Lublinerman. "Continuity & Robustness of Programs". In: *Communications of the ACM* 55.8 (2012), pp. 107–115.

[65] N. Chomsky. "Transformational Analysis". PhD thesis. University of Pennsylvania, 1955.

[66] N. Chomsky. "Three models for the description of language". In: *I.R.E. Trans. on Information Theory* IT-2 (1956), pp. 113–123.

[67] N. Chomsky. *Syntactic Structures*. The Hague/Paris: Mouton, 1957.

[68] N. Chomsky. "On certain formal properties of grammars". In: *Information and Control* 2 (1959), pp. 137–167.

[69] A. Church. "A set of postulates for the foundation of logic". In: *Annals of Mathematics* 2 (1932-33), pp. 33–34, 346–366, 839–864.

[70] F. Cohen. "Computer Viruses: Theory and Experiments". In: *Computers and Security* 6.1 (1987), pp. 22–35.

[71] I. Bernard Cohen. *Howard Aiken: Portrait of a Computer Pioneer.* MIT Press, 1999.

[72] T.R. Colburn. *Philosophy and Computer Science.* Ed. by J.H. Fetzer. M.E. Sharpe, 2000.

[73] A. Colmerauer and P. Roussel. "The birth of Prolog". In: *History of Programming Languages – II.* Ed. by T.J. Bergin Jr. and R.G. Gibson Jr. ACM Press, 1996, pp. 331–367.

[74] B. Cook, A. Podelski, and A. Rybalchenko. "Proving Program Termination". In: *Communications of the ACM* 54.5 (May 2011), pp. 88–98.

[75] S.B. Cooper. "Incomputability after Alan Turing". In: *Notices of the AMS* 59.6 (2012), pp. 776–784.

[76] B. Jack Copeland. *Turing: Pioneer of the Information Age.* Oxford, UK: Oxford University Press, 2012.

[77] H.B. Curry. *On the Composition of Programs for Automatic Computing.* Memorandum 9806. Silver Spring, Maryland: Naval Ordnance Laboratory, 1949.

[78] H.B. Curry and R. Feys. *Combinatory Logic. Volume I.* Amsterdam: North-Holland, 1958.

[79] U. Daepp and P. Gorkin. *Reading, Writing and Proving: a Closer Look at Mathematics.* Springer, 2003.

[80] U. Daepp and P. Gorkin. *Reading,Writing and Proving: a Closer Look at Mathematics (2nd. ed.)* Springer, 2011.

[81] O.-J. Dahl, E.W. Dijkstra, and C.A.R. Hoare. *Structured Programming.* London/New York: Academic Press, 1972.

[82] O.-J. Dahl, B. Myhrhaug, and K. Nygaard. *The SIMULA 67 common base language.* Tech. rep. Norwegian Computing Center, Oslo, 1968.

[83] O.-J. Dahl and K. Nygaard. "SIMULA—an ALGOL-based simulation language". In: *Communications of the ACM* 9.9 (1966), pp. 671–678.

[84] A. Dasgupta. *Set Theory.* Birkhäuser, 2014.

[85] S. Dasgupta. *Computer Science: A Very Short Introdcution.* Oxford University Press, 2016.

[86] M. Davis. *Computability and Unsolvability.* New York, USA: McGraw-Hill, 1958.

[87] M. Davis. "Mathematical Logic and the Origin of Modern Computers". In: *The Universal Turing Machine - A Half-Century Survey*. Ed. by R. Herken. Originally in: Studies in the History of Mathematics. Mathematical Association of America, 1987, pages 137-165. Wien: Springer, 1988, pp. 135–158.

[88] M. Davis. *Engines of Logic: Mathematicians and the origins of the Computer*. first. New York NY: W.W. Norton & Company, 2000.

[89] M. Davis. *The Universal Computer: The Road from Leibniz to Turing*. first. Florida: Norton, 2000.

[90] M. Davis. *The Universal Computer: The Road from Leibniz to Turing*. second. CRC Press, 2012.

[91] M. Davis. *Universality is Ubiquitous*. YouTube: www.youtube.com/ watch?v=ZVTgtODX0Nc. Presentation at the Turing Centennial Celebration at Princeton, 10–12 May 2012. 2012.

[92] M. Davis, R. Sigal, and E.J. Weyuker. *Computability, Complexity, and Languages: Fundamentals of Theoretical Computer Science*. second. Morgan Kaufmann, 1994.

[93] E.G. Daylight. *Interview with Van der Poel in February 2010, conducted by Gerard Alberts, David Nofre, Karel Van Oudheusden, and Jelske Schaap*. Tech. rep. 2010.

[94] E.G. Daylight. "Dijkstra's Rallying Cry for Generalization: the Advent of the Recursive Procedure, late 1950s – early 1960s". In: *The Computer Journal* 54.11 (2011), pp. 1756–1772.

[95] E.G. Daylight. *Interview with Blaauw on 29 November 2011, conducted by Gerard Alberts and Karel Van Oudheusden*. Tech. rep. 2011.

[96] E.G. Daylight. *Pluralism in Software Engineering: Turing Award Winner Peter Naur Explains*. Ed. by K. De Grave. Heverlee: Lonely Scholar, 2011.

[97] E.G. Daylight. "A Hard Look at George Dyson's Book "Turing's Cathedral: the Origins of the Digital Universe"". In: *Turing in Context II*. Lecture available on video: www.dijkstrascry.com/ presentations. 2012.

[98] E.G. Daylight. *The Dawn of Software Engineering: from Turing to Dijkstra*. Ed. by K. De Grave. See: www.lonelyscholar.com. Heverlee: Lonely Scholar, 2012.

[99] E.G. Daylight. *On Mahoney's Accounts of Turing*. 2013.

[100] E.G. Daylight. *Turing's 1936 Paper and the First Dutch Computers*. Communications of the ACM. 2013.

[101] E.G. Daylight. "A Turing Tale". In: *Communications of the ACM* 57.10 (2014), pp. 36–38.

[102] E.G. Daylight. "Towards a Historical Notion of 'Turing — the Father of Computer Science'". In: *History and Philosophy of Logic* 36.3 (2015), pp. 205–228.

[103] E.G. Daylight and S. Nanz, eds. *The Future of Software Engineering: Panel discussions, 22–23 November 2010, ETH Zurich*. Conversations. www.lonelyscholar.com. Heverlee: Lonely Scholar, Oct. 2011.

[104] E.G. Daylight, A. Vandecappelle, and F. Catthoor. "The Formalism Underlying EASYMAP: a Precompiler for Refinement-Based Exploration of Hierarchical Data Organizations". In: *Science of Computer Programming* 72.3 (Aug. 2008), pp. 71–135.

[105] T.J. Dekker, E.W. Dijkstra, and A. van Wijngaarden. *Cursus programmeren voor automatische rekenmachines*. Tech. rep. Amsterdam: MCR:CR-9, 1957.

[106] E.W. Dijkstra. *Functionele beschrijving van de ARRA*. Tech. rep. MR 12. Mathematisch Centrum Amsterdam, 1953.

[107] E.W. Dijkstra. *Communication with an Automatic Computer*. Academisch Proefschrift. Universiteit van Amsterdam, Oct. 1959.

[108] E.W. Dijkstra. "An Attempt to Unify Constituent Concepts of Serial Program Execution". In: *Proceedings of the Symposium Symbolic Languages in Data Processing*. London/New York: Gordon and Breach Science Publishers, 1962, pp. 237–251.

[109] E.W. Dijkstra. "*GOTO* Statement Considered Harmful". In: *Letters to the Editor, Communications of the ACM* 11 (1968), pp. 147–148.

[110] E.W. Dijkstra. *EWD 249: Notes on structured programming*. Tech. rep. Technische Hogeschool Eindhoven, 1969.

[111] E.W. Dijkstra. *Notes on Structured Programming*. Tech. rep. T.H.-Report 70-WSK-03. Second edition. Published as Chapter 1 of [81]. Technische Hogeschool Eindhoven, Apr. 1970.

[112] E.W. Dijkstra. *EWD 372: A simple axiomatic basis for programming language constructs*. Tech. rep. 1973.

[113] E.W. Dijkstra. *A Discipline of Programming*. Englewood Cliffs, N.J.: Prentice-Hall, 1976.

[114] E.W. Dijkstra. *EWD 1166: From my Life*. Tech. rep. The University of Texas at Austin, 1993.

[115] P.J. van Donselaar. *De ontwikkeling van electronische rekenmachines in Nederland; een historisch overzicht van Nederlandse computers*. Tech. rep. Amsterdam: Rapport Stichting Studiecentrum voor Administratieve Automatisering en Bestuurlijke Informatieverwerking, 1967.

[116]   G. Dowek. *Computation, Proof, Machine: Mathematics Enters a New Age*. Cambridge University Press, 2015.

[117]   G. Dyson. *Turing's Cathedral: The Origins of the Digital Universe*. London: Penguin Books, 2012.

[118]   E.A. Emerson. "25 Years of Model Checking". In: *The Beginning of Model Checking: A Personal Perspective*. Springer, 2008, pp. 27–45.

[119]   U. Erlingsson, Y. Younan, and F. Piessens. "Low-Level Software Security by Example". In: *Handbook of Information and Communication Security*. Ed. by P. Stavroulakis and M. Stamp. Springer, 2010, pp. 633–658.

[120]   G.R. Exner. *An Accompaniment to Higher Mathematics*. Springer, 1997.

[121]   A.D. Falkoff and K.E. Iverson. *APL\360: User's Manual, IBM, 1968*.

[122]   A.D. Falkoff and K.E. Iverson. "The Evolution of APL". In: *ACM SIGPLAN Notices* 13 (1978), pp. 47–57.

[123]   D.J. Farber, R.E. Griswold, and I.P. Polonsky. "SNOBOL, A String Manipulation Language". In: *Journal of the ACM* 11 (1964), pp. 21–30.

[124]   M. Fernández. *Models of Computation: An Introdcution to Computability Theory*. London: Springer, 2009.

[125]   J.H. Fetzer. "Program Verification: The Very Idea". In: *Communications of the ACM* 31.9 (1988), pp. 1048–1063.

[126]   E. Filiol. *Computer Viruses: from theory to applications*. Springer, 2005.

[127]   A.C. Fleck. "Isomorphism groups of automata". In: *Journal of the ACM* 9 (1962), pp. 469–476.

[128]   A.C. Fleck. "Algebraic structure of automata". PhD thesis. Michigan State University Library, 108 286 THS, 1964.

[129]   A.C. Fleck. "On the automorphism group of an automaton". In: *Journal of the ACM* 12 (1965), pp. 566–569.

[130]   A.C. Fleck. "Towards a theory of data structures". In: *Journal of Computer and System Sciences* 5 (1971), pp. 475–488.

[131]   A.C. Fleck. "On the impossibility of content exchange through the by-name parameter transmission mechanism". In: *SIGPLAN Notices* 11 (1976), pp. 38–41.

[132]   A.C. Fleck. "Formal models for string patterns". In: *Current Trends in Programming Methodology, Volume IV: Data Structuring*. Ed. by R. Yeh. Prentice-Hall, 1978, pp. 216–240.

[133]   A.C. Fleck. "Verifying abstract data types with SNOBOL4". In: *Software – Practice and Experience* 12 (1982), pp. 627–640.

[134] A.C. Fleck. "A proposal for the comparison of types in Pascal and associated semantic models". In: *Computer Languages* 9.2 (1984), pp. 71–87.

[135] A.C. Fleck. "Structuring FP-style functional programs". In: *Computer Languages* 11 (1986), pp. 55–63.

[136] A.C. Fleck. "A case study comparison of four declarative programming languages". In: *Software – Practice and Experience* 20 (1990), pp. 49–66.

[137] A.C. Fleck. "Specifying and proving object-oriented programs". In: *2004 Hawaii International Conference on Computer Sciences*. 2004, pp. 190–206.

[138] A.C. Fleck. "Prolog as the first programming language". In: *ACM SIGCSE Bulletin* 39 (2007), pp. 61–64.

[139] A.C. Fleck and R.S. Limaye. "Formal semantics and abstract properties of string pattern operations and extended formal language description mechanisms". In: *SIAM Journal on Computing* 12 (1983), pp. 166–188.

[140] T.M. Flett. *Mathematical Analysis*. McGraw-Hill, 1966.

[141] R.W. Floyd. "The syntax of programming languages—A survey". In: *IEEE Transactions on Electronic Computers* EC-13.4 (1964), pp. 346–353.

[142] M. Franssen, G. Lokhorst, and I. Poel. "Philosophy of technology". In: *Stanford Encyclopedia of Philosophy*. http://plato.stanford.edu/entries/technology, 2009.

[143] D.D. Freydenberger. "Extended regular expressions: succinctness and decidability". In: *Theory of Computing Systems* 53 (2013), pp. 159–193.

[144] J.E.F. Friedl. *Mastering Regular Expressions*. 3rd ed. O'Reilly Media, Inc., 2006.

[145] R. Frigg. "Models and fiction". In: *Synthese* 172 (2009), pp. 251–268.

[146] M.R. Garey and D.S. Johnson. *Computers and Intractability: A Guide to the Theory of NP-Completeness*. W.H. Freeman and Company, 1979.

[147] R. Garnier and J. Taylor. *Discrete Mathematics — Proofs, Structures and Applications*. CRC Press, 2010.

[148] L. Gerstein. *Introduction to Mathematical Structures and Proofs (2nd. ed.)* Springer, 2012.

[149] L. Gilbert and J. Gilbert. *Elements of Modern Algebra (7th. ed.)* Cengage Learning, 2008.

[150] S. Ginsburg. "On the reduction of superfluous states in sequential machines". In: *Journal of the ACM* 6 (1959), pp. 259–282.

[151] S. Ginsburg and H. Gordon Rice. "Two Families of Languages Related to ALGOL". In: *Journal of the ACM* 9.3 (1962), pp. 350–371.

[152] V.E. Giuliano and A.G. Oettinger. "Research on automatic translation at the Harvard Computation Laboratory". In: *Information processing: proceedings of the International Conference on Information Processing*. Unesco, Paris. 1959.

[153] J.A. Goguen. "Some design principles and theory for OBJ-0, a language for expressing and executing algebraic specifications of programs". In: *Mathematical Studies of Information Processing*. Ed. by E. Blum, M. Paul, and S. Takasu. LNCS V.75. Springer-Verlag, 1979, pp. 425–473.

[154] J.A. Goguen, J.W. Thatcher, E.G. Wagner, and J.B. Wright. "Abstract data-types as initial algebras and correctness of data representations". In: *Computer Graphics, Pattern Recognition and Data Structure*. 1975, pp. 89–93.

[155] A. Goldberg and D. Robson. *Smalltalk-80: the language and its implementation*. Addison-Wesley, 1983.

[156] E.G. Goodaire and M.M. Parmenter. *Discrete Mathematics with Graph Theory (3rd. ed.)* Pearson Prentice Hall, 2006.

[157] R. Goodman, ed. *Annual Review in Automatic Programming I: Papers read at the Working Conference on Automatic Programming of Digital Computers held at Brighton, 1–3 April 1959*. New York, USA: Pergamon Press, 1960.

[158] S. Gorn. "Planning Universal Semi-Automatic Coding". In: *Symposium on Automatic Programming for Digital Computers*. Office of Naval Research, Department of the Navy. Washington D.C., May 1954, pp. 74–83.

[159] S. Gorn. *Real Solutions Of Numerical Equations By High Speed Machines*. Tech. rep. 966. Available from the "Saul Gorn Papers" from the University of Pennsylvania Archives (unprocessed collection). Ballistic Research Laboratories, 1955.

[160] S. Gorn. "Standardized Programming Methods and Universal Coding". In: *Journal of the ACM* (July 1957). Received in December 1956.

[161] S. Gorn. *Common Programming Language Task, Final Report: Report of the work in the period 1 May 1958 to 30 June 1959*. Tech. rep. AD59UR1. Available from the "Saul Gorn Papers" from the University of Pennsylvania Archives (unprocessed collection): UPT 50 G671 Box 39. 1959.

[162] S. Gorn and W. Manheimer. *The electronic brain and what it can do.*
       Ed. by P.F. Brandwein. Chicago, Illinois, USA: Science Research
       Associates, Inc., 1956.

[163] E.M. Grabbe, S. Ramo, and D.E. Wooldridge, eds. *Handbook
       of Automation, Computation, and Control: Computers and Data
       Processing.* Vol. 2. New York: John Wiley and Sons, Inc., 1959.

[164] D. Gries and F.B. Schneider. *A Logical Approach to Discrete Math.*
       Springer, 1993.

[165] R.E. Griswold. "History of Programming Languages". In: ACM
       Press, 1981. Chap. A history of the SNOBOL programming
       languages, pp. 601–645.

[166] R.E. Griswold, J.F. Poage, and I.P. Polonsky. *The SNOBOL4
       Programming Language.* Prentice-Hall, 1968.

[167] R.E. Griswold, J.F. Poage, and I.P. Polonsky. *The SNOBOL4
       Programming Language.* 2nd ed. Prentice-Hall, 1971.

[168] J.V. Guttag, E. Horowitz, and D.R. Musser. "The design of data
       type specifications". In: *ICSE'76: 2nd International Conference on
       Software Engineering.* IEEE, 1976, pp. 414–430.

[169] J.V. Guttag, E. Horowitz, and D.R. Musser. "Abstract data types
       and software validation". In: *Communications of the ACM* 21 (1978),
       pp. 1048–1063.

[170] T. Haigh, M. Priestley, and C. Rope. *ENIAC in Action.* MIT Press,
       2016, p. 360.

[171] P.R. Halmos. "Nicolas Bourbaki". In: *Scientific American* 196.5
       (1957), pp. 88–99.

[172] P.R. Halmos. *Naive Set Theory.* Van Nostrand Reinhold, 1960.

[173] R. Hammack. *Book of Proof.* CC BY-ND, 2009.

[174] D. Harel. *The Science of Computing: Exploring the Nature and Power
       of Algorithms.* Addison-Wesley, 1987, 1989.

[175] I.N. Herstein. *Topics in Algebra.* Xerox College Publishing, 1964.

[176] M. Hicks. *How did Heartbleed remain undiscovered, and what
       should we do about it?* This post is dated "July 1st, 2014" and I
       accessed it on September 19th, 2016. URL: http://www.pl-
       enthusiast.net/2014/07/01/how-did-heartbleed-remain-
       undiscovered-and-what-should-we-do-about-it/.

[177] C.A.R. Hoare. "Record handling". In: *ALGOL Bulletin* 21 (1965),
       pp. 39–69.

[178] C.A.R. Hoare. "An Axiomatic Basis for Computer Programming".
       In: *Communications of the ACM* 12.10 (1969), pp. 576–580.

[179] C.A.R. Hoare. "Proof of correctness of data representations". In: *Acta Informatica* 1 (1972), pp. 271–281.

[180] C.A.R. Hoare. "The Emperor's Old Clothes". In: *Communications of the ACM* 24.2 (1981), pp. 75–83.

[181] C.A.R. Hoare and D.C.S. Allison. "Incomputability". In: *ACM Computing Surveys* 4.3 (1972), pp. 169–178.

[182] A. Hodges. *Alan Turing: The Enigma*. London: Burnett Books, 1983.

[183] A. Hodges. "Book Review: The Essential Turing". In: *Notices of the AMS* 53.10 (2006), pp. 1190–1199.

[184] G. van den Hove. "On the Origin of Recursive Procedures". In: *The Computer Journal* 58.11 (2015), pp. 2892–2899.

[185] R.K.W. Hui, K.E. Iverson, E.E. McDonnell, and A.T. Whitney. "APL\?" In: *APL1990*. 1990, pp. 192–200.

[186] Yu. I. Ianov. "On the equivalence and transformation of program schemes". In: *Doklady Aka. Nauk S.S.S.R.* 113 (1957), pp. 39–42.

[187] *IBM, 650 Magnetic Drum Data-Processing Machine Manual of Operation, Form 22-6060-1, 1955.*

[188] *IBM, Programmer's Reference Manual (for the) Fortran Automatic Coding System for the IBM704, 1956.*

[189] *IEEE Standard Pascal Computer Programming Language, ANSI/IEEE X3.97-1983*. American National Standards Institute, Inc., 1983.

[190] N. Irmak. "Software is an Abstract Artifact". In: *Grazer Philosophische Studien* 86 (2012), pp. 55–72.

[191] ISO/IEC. *Quantities and units — Part 2: Mathematical signs and symbols to be used in the natural sciences and technology (ISO 80000-2)*. ISO/IEC, 2009.

[192] K.E. Iverson. *A Programming Language*. New York: John Wiley and Sons, Inc., 1962.

[193] M.A. Jackson and E.G. Daylight. *Formalism and Intuition in Software Development*. Ed. by K. De Grave. Geel: Lonely Scholar, 2015.

[194] T. Jech. *Set Theory*. Springer, 2003.

[195] K. Jensen and N. Wirth. *Pascal User Manual and Report*. Springer-Verlag, 1974.

[196] R.T. Johnson and J.B. Morris. "Abstract data types in the MODEL programming language". In: *SIGPLAN Notices* 11 (1976), pp. 36–46.

[197] A.A. Markov Jr. *Theory of Algorithms*. Vol. 42. Trudy Matematicheskogo Instituta imeni V. A. Steklova. Moscow/Leningrad: Academy of Sciences of the USSR, 1954.

[198] *Jsoftware Inc., http://www.jsoftware.com*.

[199] D. Kahn. *The Codebreakers: The Story of Secret Writing*. New York: The Macmillan Company, 1967.

[200] L.V. Kantorovich. "On a mathematical symbolism convenient for performing machine calculations". In: *Doklady Aka. Nauk S.S.S.R.* 113 (1957), pp. 738–741.

[201] A.C. Kay. "The early history of Smalltalk". In: *History of Programming Languages – II*. Ed. by T.J. Bergin Jr. and R.G. Gibson Jr. ACM Press, 1996, pp. 511–598.

[202] S.C. Kleene. *Introduction to Metamathematics*. Princeton, New Jersey, USA: Van Nostrand, 1952.

[203] S.C. Kleene. "Representation of events in nerve nets and finite automata". In: *Automata Studies*. Ed. by C.E. Shannon and J. McCarthy. Princeton University Press, 1956, pp. 3–42.

[204] G. Klein et al. "seL4: Formal Verification: of an OS Kernel". In: *SOSP*. 2009.

[205] D.E. Knuth. "The remaining trouble spots in ALGOL 60". In: *Communications of the ACM* 10 (1967), pp. 611–618. Reprinted with corrections and an addendum in [209].

[206] D.E. Knuth. "Semantics of Context-Free Languages". In: *Mathematical Systems Theory* 2 (1968), pp. 127–145. Reprinted with corrections and an addendum in [209].

[207] D.E. Knuth. "Literate Programming". In: *The Computer Journal* 27 (1984), pp. 97–111. Reprinted as Chapter 4 of [208].

[208] D.E. Knuth. *Literate Programming*. Vol. 27. CSLI Lecture Notes. Stanford, California: CSLI Publications, 1992.

[209] D.E. Knuth. *Selected Papers on Computer Languages*. Vol. 139. CSLI Lecture Notes. Stanford, California: CSLI Publications, 2003.

[210] D.E. Knuth and E.G. Daylight. *Algorithmic Barriers Falling: P=NP?* Ed. by K. De Grave. Geel: Lonely Scholar, 2014.

[211] A.L. Kolmogorov and S.V. Fomin. *Introductory Real Analysis*. Dover, 1970.

[212] E. Kranakis. "Early Computers in The Netherlands". In: *CWI-Quarterly* (), pp. 61–274.

[213] S.G. Krantz. *Real Analysis and Foundations*. Chapman & Hall/CRC, 2005.

[214]   D. Kroening and O. Strichman. *Decision Procedures: An Algorithmic Point of View*. Springer, 2008.

[215]   P. Kroes. "Engineering and the dual nature of technical artefacts". In: *Cambridge Journal of Economics* 34 (2010), pp. 51–62.

[216]   L. Lamport. *Specifying Systems — The TLA+ Language and Tools for Hardware and Software Engineers*. Pearson, 2003.

[217]   L. Lamport and L.C. Paulson. "Should your specification language be typed?" In: *ACM TOPLAS* 21.3 (1999), pp. 502–526.

[218]   S. Lang. *Undergraduate Analysis*. Springer, 1983.

[219]   R. Larson and B. Edwards. *Calculus 9e*. Brooks/Cole, 2009.

[220]   S. Lavington, ed. *Alan Turing and his Contemporaries: Building the world's first computers*. Swindon, UK: bcs, 2012.

[221]   D. Leavitt. *The Man Who Knew Too Much: Alan Turing and the Invention of the Computer*. New York: Atlas Books, 2006.

[222]   D. Leavitt. *Alan Turing, l'homme qui inventa l'informatique*. Paris: Dunod, 2007.

[223]   E.A. Lee. "Absolutely positive on time: what would it take?" In: *Computer* 38.7 (2005), pp. 85–87.

[224]   E.A. Lee and P. Varaiya. *Structure and Interpretation of Signals and Systems*. Addison Wesley / Pearson Education, 2003.

[225]   J.A.N. Lee. *Computer pioneers*. California: IEEE Computer Society Press, 1995.

[226]   H. Lewis and C. Papadimitriou. *Elements of the Theory of Computation*. Ed. by Engelwood Cliffs. New Jersey: Prentice-Hall, 1981.

[227]   B. Liskov and S. Zilles. "Programming with abstract data types". In: *ACM SIGPLAN Notices* 9 (1974), pp. 50–59.

[228]   B.H. Liskov and S. Zilles. "Specification Techniques for Data Abstractions". In: *IEEE Transactions on Software Engineering* (1975), pp. 72–87.

[229]   K.C. Liu and A.C. Fleck. "String pattern matching in polynomial time". In: *6th ACM Symposium on Principles of Programming Languages*. 1979, pp. 222–225.

[230]   W.N. Locke and A.D. Booth, eds. *Machine Translation of Languages: Fourteen Essays*. Cambridge, MA and New York: MIT Press and John Wiley & Sons, Inc., 1955.

[231]   R.L. London. *Who Earned First Computer Science Ph.D.?* BLOG @CACM (2013), http://cacm.acm.org/blogs/blog-cacm/159591-who-earned-first-computer-science-ph-d/fulltext.

[232] S. Mac Lane. *Categories for the Working Mathematician*. Springer, 1971.

[233] D. MacKenzie. *Mechanizing Proof: Computing, Risk, and Trust*. MIT Press, 2004.

[234] M.S. Mahoney. *Histories of Computing*. Ed. by T. Haigh. Cambridge, Massachusetts/London, England: Harvard University Press, 2011.

[235] Z. Manna and J. Vuillemin. "Fixpoint approach to the theory of computation". In: *Communications of the ACM* 15 (1972), pp. 528–536.

[236] J. Martin-Nielsen. "'This war for men's minds': the birth of a human science in Cold War America". In: *History of the Human Sciences* 23.5 (2010), pp. 131–155.

[237] S. Martini. "Several Types of Types in Programming Languages". In: *History and Philosophy of Computing 2015*. 2015, pp. 216–227.

[238] J. McCarthy. "Recursive Functions of Symbolic Expressions and Their Computation by Machine, Part I". In: *Communications of the ACM* 3.4 (1960), pp. 184–195.

[239] J. McCarthy. "History of Programming Languages". In: ed. by R.L. Wexelblat. New York: Academic Press, 1981. Chap. 'History of LISP' and the transcripts of: presentation, discussant's remark, question and answer session, pp. 173–195.

[240] W.S. McCulloch and W. Pitts. "A Logical Calculus of the Ideas Immanent in Nervous Activity". In: *Bull. math. Biophys.* 5 (1943), pp. 115–133.

[241] G.H. Mealy. "A method for synthesizing sequential circuits". In: *Bell Systems Technology Journal* 34 (1955), pp. 1045–1079.

[242] E. Mendelson. *Introduction to Mathematical Logic (3rd. ed.)* Wadsworth & Brooks / Cole, 1987.

[243] A.R. Meyer and D.M. Ritchie. "The complexity of loop programs". In: *Proceedings of the ACM National Meeting*. 1967, pp. 465–469.

[244] B. Meyer. *Introduction to the Theory of Programming Languages*. Prentice Hall, 1991.

[245] M. Minsky. *Computation: Finite and Infinite Machines*. Prentice-Hall, Inc, 1967.

[246] L. De Mol. "Doing mathematics on the ENIAC. Von Neumann's and Lehmer's different Visions." In: *Mathematical practice and development throughout History*. Ed. by E. Wilhelmus and I. Witzke. Logos Verlag, Berlin, 2009, pp. 149–186.

[247] L. De Mol and M. Bullynck. "A short history of small machines". In: *CiE 2012 - How the World Computes*. 2012.

[248] L. De Mol, M. Bullynck, and M. Carle. "Haskell before Haskell. Curry's contribution to a theory of programming." In: *Programs, Proofs, Processes, Computability in Europe 2010*. Vol. 6158. LNCS. 2010, pp. 108–117.

[249] E.F. Moore. "Gedanken-experiments on sequential machines". In: *Automata Studies*. Princeton University Press, 1956, pp. 129–153.

[250] H.L. Morgan and R.A. Wagner. "PL/C: the design of a high-performance compiler for PL/I". In: *AFIPS Spring Joint Computing Conference*. 1971, pp. 503–510.

[251] P. Mounier-Kuhn. "Comment l'informatique devint une science". In: *La Recherche* 465 (2012), pp. 92–94.

[252] P. Mounier-Kuhn. "Computer Science in French Universities: Early Entrants and Latecomers". In: *Information & Culture: A Journal of History* 47.4 (2012), pp. 414–456.

[253] P. Mounier-Kuhn. "Logic and Computing in France: A Late Convergence". In: *AISB/IACAP World Congress 2012 — History and Philosophy of Programming*. Ed. by L. De Mol and G. Primiero. 2012, pp. 44–47.

[254] P. Mounier-Kuhn. "Algol in France: From Universal Project to Embedded Culture". In: *IEEE Annals of the History of Computing* 36.4 (2014), pp. 6–25.

[255] S. Nanz, ed. *The Futurue of Software Engineering*. Springer, 2011.

[256] P. Naur. "The European side of the last phase of the development of ALGOL 60". In: *History of Programming Languages*. Ed. by R.L. Wexelblat. New York: Academic Press, 1981, pp. 92–139.

[257] P. Naur. *Computing: A Human Activity*. New York: ACM Press / Addison-Wesley, 1992.

[258] A. Newell and F.M. Tonge. "An Introduction to Information Processing Language V". In: *Communications of the ACM* 3.4 (1960), pp. 205–211.

[259] D. Nofre. "Unraveling Algol: US, Europe, and the Creation of a Programming Language". In: *IEEE Annals of the History of Computing* 32.2 (2010), pp. 58–68.

[260] D. Nofre. *Alan Jay Perlis*. Short bio of Perlis, available on the official ACM website: amturing.acm.org/award_winners. 2012.

[261] D. Nofre, M. Priestley, and G. Alberts. "When Technology Became Language: The Origins of the Linguistic Conception of Computer Programming, 1950–1960". In: *Technology and Culture* 55.1 (2014), pp. 40–75.

[262] K.V. Nori, U. Ammann, K. Jensen, H.H. Nageli, and C. Jacobi. "Pascal-P Implementation Notes". In: *Pascal – The Language and*

its Implementation. Ed. by D.W. Barron. John Wiley and Sons, Inc., 1981, pp. 125–170.

[263] M.J. O'Donnell. "Computing in Systems Described by Equations". In: *Lecture Notes in Computer Science* 58 (1977).

[264] A.G. Oettinger. "Programming a Digital Computer to Learn". In: *Philosophical Magazine* 43.347 (1952), pp. 1243–1263.

[265] A.G. Oettinger. "Account identification for Automatic Data Processing". In: *Journal of the ACM* 4.3 (1957), pp. 245–253.

[266] A.G. Oettinger. *Automatic Language Translation*. Cambridge, Massachusetts, USA: Harvard University Press, 1960.

[267] A.G. Oettinger. "Automatic Syntactic Analysis and the Pushdown Store". In: *Proceedings of Symposium in Applied Mathematics*. Vol. 12. Providence: American Mathematical Society, 1961, pp. 104–129.

[268] A.G. Oettinger. "Reminiscences of the Boss". In: *Makin' Numbers: Howard Aiken and the Computer*. Ed. by I.B. Cohen and G.W. Welch. Cambridge, Massachusetts, USA: MIT Press, 1999, pp. 203–214.

[269] A. Olley. "Existence Precedes Essence — Meaning of the Stored-Program Concept". In: *IFIP Advances in Information and Communication Technology*. Ed. by A. Tatnall. Vol. 325. Springer, 2010, pp. 169–178.

[270] C.H. Papadimitriou. *The Origin of Computable Numbers: A Tale of Two Classics*. YouTube: www.youtube.com/watch?v=IZWjFCTx0 OQ. Presentation at the Turing Centennial Celebration at Princeton, 10–12 May 2012. 2012.

[271] D.L. Parnas. "On Proving Continuity of Programs". In: *Letter to the Editor of the Communications of the ACM* 55.11 (2012), p. 9.

[272] A.J. Perlis. "Announcement". In: *Communications of the ACM* 1.1 (1958).

[273] B.C. Pierce. *Basic Category Theory for Computer Scientists*. The MIT Press, 1991.

[274] W. van der Poel. *Inzending 1946/47 van Van der Poel op de prijsvraag genaamd "1+1=10"*. Tech. rep. Delft, 1948.

[275] W.L. van der Poel. "A Simple Electronic Digital Computer". In: *Appl. sci. Res.* 2 (1952), pp. 367–399.

[276] W.L. van der Poel. "The Logical Principles of Some Simple Computers". PhD thesis. Universiteit van Amsterdam, 1956.

[277] W.L. van der Poel. "Digitale Informationswandler". In: ed. by W. Hoffmann. Braunschweig: Vieweg, 1961. Chap. Microprogramming and trickology, pp. 269–311.

[278] W.L. van der Poel. *Een leven met computers*. TU Delft. Oct. 1988.

[279]   *Preliminary Report – Specifications for the IBM Mathematical FORmula TRANslating System.* Tech. rep. New York: IBM, Programming Research Group, 1954.

[280]   M. Priestley. *A Science of Operations: Machines, Logic and the Invention of Programming.* Ed. by M. Campbell-Kelly. London: Springer, 2011.

[281]   P.M. Priestley. *Logic and the Development of Programming Languages, 1930-1975.* PhD thesis. University College London, May 2008.

[282]   W.V. Quine. *Set Theory and Its Logic.* The Belknap Press of Harvard University Press, 1969.

[283]   M.O. Rabin and D. Scott. "Finite Automata and their Decision Problems". In: *IBM Journal of Research and Development* 3.2 (1959), pp. 114–125.

[284]   G. Radin and H.P. Rogoway. "NPL: Highlights of a New Programming Language". In: *Communications of the ACM* 8 (1965), pp. 9–17.

[285]   B. Randell, ed. *The Origins of Digital Computers: Selected Papers.* Berlin Heidelberg New York: Springer-Verlag, 1973.

[286]   B. Randell and L.J. Russell. *ALGOL60 Implementation.* Academic Press, 1964.

[287]   S.N. Razumovskii. "On the question of automatization of programming of problems of translation from one language to another". In: *Doklady Aka. Nauk S.S.S.R.* 113 (1957), pp. 760–762.

[288]   J.C. Reynolds. *Letter to the author on 9 March 2012.*

[289]   C.J. van Rijsbergen. "Turing and the origins of digital computers". In: *Aslib Proceedings.* Vol. 37. 6/7. Paper presented at an Aslib Evening Meeting, Aslib, Information House, 27 March 1985. Emerald Backfiles, 1985, pp. 281–285.

[290]   C.E. Jr. Roberts. *Introduction to Mathematical Proofs — A Transition.* CRC Press, 2010.

[291]   J.A. Robinson. "A machine-oriented logic based on the resolution principle". In: *Journal of the ACM* 12 (1965), pp. 23–41.

[292]   J.A. Robinson. "Logic, computers, Turing, and von Neumann". In: *Machine Intelligence 13: Machine Intelligence and Inductive Learning.* Ed. by K. Furukawa, D. Michie, and S. Muggleton. Oxford, UK: Clarendon Press, 1994, pp. 1–35.

[293]   C. Rooijendijk. *Alles moest nog worden uitgevonden: De geschiedenis van de computer in Nederland.* Olympus, 2007.

[294]   P. Rosenbloom. *Elements of Mathematical Logic.* New York: Dover, 1950.

[295]   H.L. Royden. *Real Analysis.* Macmillan, 1968.

[296] W. Rudin. *Principles of Mathematical Analysis*. McGraw-Hill, 1964.

[297] H. Rutishauser. "The Use of Recursive Procedures". In: *Annual Review in Automatic Programming 3*. Ed. by R. Goodman. New York: Pergamon Press, 1963, pp. 43–52.

[298] J.G. Sanderson. "On Simple Low Redundancy Languages". In: *Communications of the ACM* 8.10 (1965). Letters to the Editor.

[299] S.R. Sataluri and A.C. Fleck. "Semantic specification using logic programs". In: *Logic Programming, Proceedings of the North American Conference 1989*. Ed. by E.L. Lusk and R.A. Overbeek. Vol. 2. Cleveland, Ohio: MIT Press, 1989, pp. 772–794.

[300] E.R. Scheinerman. *Mathematics — A Discrete Introduction (3rd. ed.)* Cengage Learning, 2012.

[301] E. Shapiro. "Separating Concurrent Languages with Categories of Language Embeddings". In: *Proceedings of the Twenty-third Annual ACM Symposium on Theory of Computing*. 198-208. 1991.

[302] H.S. Shuard. "Does it matter?" In: *The Mathematical Gazette* 59.407 (1976), pp. 7–15.

[303] B.C. Smith. "The Limits of Correctness". In: *ACM SIGCAS Computers and Society* 14,15 (1985), pp. 18–26.

[304] D. Smith, M. Eggen, and R. St Andre. *A Transition to Advanced Mathematics*. Cengage Learning, 2010.

[305] J.M. Spivey. *The Z Notation – A Reference Manual*. http://spivey .oriel.ox.ac.uk/~mike/zrm/. Prentice Hall, 1989.

[306] D. Sprecher. *Elements of Real Analysis*. Academic Press, 1970.

[307] J.B. Stewart. *Calculus: Early Transcendentals (7th. ed.)* Cengage Learning, 2010.

[308] H. Stoyan. "Early LISP History (1956-1959)". In: *LISP and Functional Programming*. 1984, pp. 299–310.

[309] C. Strachey. "An impossible program". In: *Letter to the Editor of the Computer Journal* (1965), p. 313.

[310] C. Strachey. *The Varieties of Programming Language*. Monograph PRG-10. Oxford University, Computing Laboratory, 1973.

[311] C. Strachey and M.V. Wilkes. "Some proposals for improving the efficiency of ALGOL 60". In: *Communications of the ACM* 4.11 (1961). Also in: University Mathematical Laboratory Technical Memorandum, No. 61/5, pp. 488–491.

[312] P. Suppes. *Axiomatic Set Theory*. Dover, 1972.

[313] I.E. Sutherland. "Sketchpad: a man-machine graphical communication system". In: *AFIPS'63 — Spring Joint Computer Conference*. 1963, pp. 392–346.

[314]   *Symposium on Advanced Programming Methods for Digital Computers — Washington, D.C., June 28,29, 1956*. Available from the "Saul Gorn Papers" from the University of Pennsylvania Archives (unprocessed collection): UPT 50 G671 Box 43. 1956.

[315]   *Symposium on Automatic Programming for Digital Computers*. Office of Naval Research, Department of the Navy. Washington D.C., May 1954.

[316]   A. Tarski and S. Givant. *A Formalization of Set Theory Without Variables*. Reprinted with corrections 1988. The American Mathematical Society, 1987.

[317]   M. Tedre. *The Science of Computing: Shaping a Discipline*. Taylor and Francis, 2014.

[318]   K. Thompson. "Regular expression search algorithm". In: *Communications of the ACM* 11.6 (1968), pp. 419–422.

[319]   H. Tuch, G. Klein, and M. Norrish. "Types, Bytes, and Separation Logic". In: *Principles of Programming Languages*. 2007.

[320]   A.M. Turing. "On Computable Numbers, with an Application to the Entscheidungsproblem". In: *Proceedings of the London Mathematical Society, 2nd series* 42 (1936), pp. 230–265. Corrections provided in [321].

[321]   A.M. Turing. "On Computable Numbers, with an Application to the Entscheidungsproblem. A Correction". In: *Proceedings of the London Mathematical Society, 2nd series* 43 (1937).

[322]   A.M. Turing. "Computing machinery and intelligence". In: *Mind* 59 (1950), pp. 433–460.

[323]   D.A. Turner. "A new implementation technique for applicative languages". In: *Software – Practice and Experience* 9 (1979), pp. 31–49.

[324]   D.A. Turner. "Miranda: a non-strict functional language with polymorphic types". In: *IFIP International Conference on Functional Programming Languages and Computer Architecture*. Vol. 201. Lecture Notes in Computer Science. Springer-Verlag, 1985, pp. 1–16.

[325]   R. Turner. "Programming Languages as Technical Artefacts". In: *Philosophy and Technology* 27.3 (2014). First online: 13 February 2013, pp. 377–397.

[326]   *UNCOL Committee Report*. From the Charles Babbage Institute collections. Thanks to David Nofre for giving me a copy of this letter. 1961.

[327]   *University of Pennsylvania Computer Activity Report: July 1, 1959 – December 31, 1960*. Tech. rep. Available from the "Saul Gorn Pa-

pers" from the University of Pennsylvania Archives (unprocessed collection): UPT 50 G671 Box 39. Office of Computer Research and Education, 1960.

[328] *USA Standard FORTRAN, ANSI X3.9-1966*. USA Standards Institute, Inc.

[329] M.Y. Vardi. "Who Begat Computing?" In: *Communications of the ACM* 56.1 (2013), p. 5.

[330] D.J. Velleman. *How To Prove It: A Structured Approach (2nd. ed.)* 5th printing. Cambridge, 2009.

[331] C.J.D.M. Verhagen. *Rekenmachines in Delft*. Uitgave van de Commissie Rekenmachines van de Technische Hogeschool te Delft. 1960.

[332] B. Bensaude Vincent. "Discipline-building in synthetic biology". In: *Studies in History and Philosophy of Science Part C: Studies in History and Philosophy of Biological and Biomedical Sciences* 44.2 (2013), pp. 122–129.

[333] J. Voeten. "On the fundamental limitations of transformational design". In: *ACM Transactions on Design Automation of Electronic Systems* 6.4 (2001), pp. 533–552.

[334] W.D. Wallis. *A Beginner's Guide to Discrete Mathematics (2nd. ed.)* Birkhäuser, 2012.

[335] R.L. Wexelblat, ed. *History of Programming Languages*. New York: Academic Press, 1981.

[336] A. van Wijngaarden. *Switching and Programming*. Tech. rep. Report MR 50. Mathematisch Centrum Amsterdam, 1962.

[337] A. van Wijngaarden. "Generalized ALGOL". In: *Annual Review in Automatic Programming*. Ed. by R. Goodman. Vol. 3. New York: Pergamon Press, 1963, pp. 17–26.

[338] A. van Wijngaarden. "Numerical analysis as an independent science". In: *BIT* 6 (1966), pp. 66–81.

[339] A. van Wijngaarden and E.W. Dijkstra. *Programmeren voor Automatische Rekenmachines*. Amsterdam, 1955.

[340] Wikipedia. *List of programming languages*. https://en.wikipedia.org/wiki/List_of_programming_languages.

[341] M.V. Wilkes. "Can Machines Think?" In: *Spectator* 6424 (1951), pp. 177–178.

[342] M.V. Wilkes. *Memories of a Computer Pioneer*. MIT Press, 1985.

[343] N. Wirth. "The programming language Pascal". In: *Acta Informatica* 1 (1971), pp. 35–63.

[344]   N. Wirth. "Recollections about the development of Pascal". In: *History of Programming Languages*. Ed. by T.J. Bergin Jr. and R.G. Gibson Jr. Addison-Wesley, 1996, pp. 97–111.

[345]   N. Wirth and C.A.R. Hoare. "A contribution to the development of ALGOL". In: *Communications of the ACM* 9 (1966), pp. 423–432.

[346]   N. Wirth and H. Weber. "EULER: A Generalization of ALGOL, and its Formal Definition: Part I". In: *Communications of the ACM* 9.1 (Jan. 1966), pp. 13–25. Part II, ibid. 9,2 (1966) 89-99.

[347]   W. Wulf, R. London, and M. Shaw. "An introduction to the construction and verification of Alphard programs". In: *IEEE Transactions on Software Engineering* 2.4 (1976), pp. 253–265.

[348]   E. Zakon. *Mathematical Analysis, Vol. I*. http://www.trillia.com /d9/zakon-analysisI-a4-one.pdf. Trillia, 2004.

Index

Also by Edgar G. Daylight:

Conversations

- Pluralism in Software Engineering:
  Turing Award Winner Peter Naur Explains
  2011 · ISBN 9789491386008

- Panel discussions I & II, held at the Future of Software
  Engineering Symposium
  2011 · ISBN 9789491386015

- The Essential Knuth
  2013 · ISBN 9789491386039

- Algorithmic Barriers Falling: P=NP?
  2014 · ISBN 9789491386046

- Formalism & Intuition in Software Development
  2015 · ISBN 9789491386053

Full-length books

- The Dawn of Software Engineering:
  from Turing to Dijkstra
  2012 · ISBN 9789491386022

Find our latest publications at `www.lonelyscholar.com`.

LONELY SCHOLAR™
SCIENTIFIC BOOKS

From the same author:

The Essential Knuth

Donald E. Knuth lived two separate lives in the late 1950s. During daylight he ran down the visible and respectable lane of mathematics. During nighttime, he trod the unpaved road of computer programming and compiler writing.

Both roads intersected! — as Knuth discovered while reading Noam Chomsky's book *Syntactic Structures* on his honeymoon in 1961.

> Chomsky's theories fascinated me, because they were mathematical yet they could also be understood with my programmer's intuition. It was very curious because otherwise, as a mathematician, I was doing integrals or maybe was learning about Fermat's number theory, but I wasn't manipulating symbols the way I did when I was writing a compiler. With Chomsky, wow, I was actually doing mathematics and computer science simultaneously.

How, when, and why did mathematics and computing converge for Knuth? To what extent did logic and Turing machines appear on his radar screen?

The early years of convergence ended with the advent of Structured Programming in the late 1960s. How did that affect his later work on TₑX? And what did "structure" come to mean to Knuth?

The Dawn of Software Engineering: from Turing to Dijkstra

The Dawn of Software Engineering is a rich and fascinating account of the time when software engineering was a compelling intellectual discipline at the center of computer science.
— John C. Reynolds, Carnegie Mellon University

Problems unsolvable with a computer influenced experienced programmers, including Edsger W. Dijkstra. His pioneering work shows that both unsolvability & aesthetics have practical relevance in software engineering. But to what extent did Dijkstra and others depend on Alan Turing's accomplishments? Daylight discusses the emerging field of software engineering with four Turing award winners: Hoare, Liskov, Wirth, and Naur. This book presents a revealing synthesis for the modern software engineer.

Wonderfully novel, very readable, and most engaging
— Grady Booch, IBM Fellow

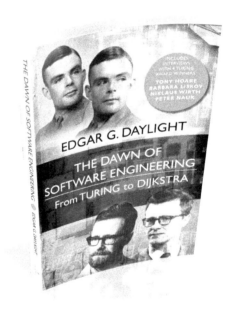

www.ingramcontent.com/pod-product-compliance
Lightning Source LLC
LaVergne TN
LVHW022305060326

832902LV00020B/3278